Acclaim for *When Science Meets Religion*

"Ian Barbour is the *doyen* of contemporary writers on science and theology. In this survey of his thinking he writes with his customary well-organized clarity."

—John Polkinghorne, author of *Belief in God in an Age of Science*

"Ian Barbour here distills a lifetime of thinking about how science and religion relate. In spanning the spectrum of the natural and the human sciences, he knows more about this than any other person on Earth, more, I suppose than any other in history. Couple this with his sure-footed capacity for balanced evaluation—a particular focus distinguishing this from his earlier books—and the result is outstanding."

—Holmes Rolston, III, author, *Genes, Genesis and God*, University Distinguished Professor, Colorado State University

"At last we have a volume in religion and science by the pioneer of the field, Ian G. Barbour, which uses Barbour's now classic four-fold typology to address fundamental issues of importance to us all. This book will be an invaluable resource to teachers, scholars, ministers, scientists and everyday inquirers who want to become part of the positive and creative interaction now growing rapidly and internationally between religion and science. Read this book and prepare for a wondrous experience!"

—Robert John Russell, Director of The Center for Theology and the Natural Sciences, Berkeley

"No surer and fairer guide to the proliferating literature on the relation of science and religion can be found than Ian Barbour. In this volume he has made accessible the fruits of his extensive and scholarly studies to those coming new to the field. They, and also those already involved in it, will welcome this elegantly organized and presented work."

—Arthur Peacocke, Oxford University, author of *From DNA to Dean:Reflections & Explorations of a Priest-Scientist*

When Science
Meets Religion

When Science Meets Religion

Ian G. Barbour

HarperSanFrancisco
A Division of HarperCollins*Publishers*

HarperCollins books may be purchased for educational, business, or sales promotional use. For information please write: Special Markets Department, HarperCollins Publishers Inc., 10 East 53rd Street, New York, NY 10022.

HarperCollins Web site: http://www.harpercollins.com

HarperCollins®, 🕮 ®, and HarperSanFrancisco™ are trademarks of HarperCollins Publishers Inc.

FIRST EDITION

Designed by Kris Tobiassen

Library of Congress Cataloging-in-Publication Data

Barbour, Ian G.
  When science meets religion / Ian G. Barbour.—1st ed.
    p. cm.
  Includes bibliographical references and index.
  ISBN 0–06–060381–X (pbk.)
  1. Religion and science. I. Title.
BL240.2 .B375 2000
291.1'75—dc21   99-055579

00 01 02 03 ❖/RRD 10 9 8 7 6 5 4 3

For my grandchildren

Graham Ojala-Barbour
Alexandra Albers
Reed Ojala-Barbour

for whom science and religion will
continue to raise significant issues
as they enter this new millennium

Contents

Preface

When religion first met modern science in the seventeenth century, the encounter was a friendly one. Most of the founders of the scientific revolution were devout Christians who held that in their scientific work they were studying the handiwork of the Creator. By the eighteenth century many scientists believed in a God who had designed the universe, but they no longer believed in a personal God actively involved in the world and human life. By the nineteenth century some scientists were hostile to religion—though Darwin himself maintained that the process of evolution (but not the details of particular species) was designed by God.

In the twentieth century the interaction of religion and science has taken many forms. New discoveries in science have challenged many classical religious ideas. In response, some people have defended traditional doctrines, others have abandoned the tradition, and still others have reformulated long-held concepts in the light of science. As we enter the new millennium, there is evidence of renewed interest in these issues among scientists, theologians, the media, and the public. Six of the most widely debated questions are taken up in successive chapters of this book.

Science and Religion: Enemies, Strangers, or Partners?

Science and religion are often seen as enemies locked in mortal combat. Some people in both camps are aggressively continuing the warfare, particularly on the topic of evolution. But conflict can be avoided if science and religion are strangers occupying separate domains at a safe distance from each other. Science asks about causal relations between events, it is said, while religion asks about the meaning and purpose of our lives. The two kinds of inquiry offer complementary perspectives on the world, separate and independent from each other and not in conflict. However, many people today are seeking a more constructive partnership. They hold that science raises questions it cannot itself answer. Why is there a universe at all? Why does it have the kind of order it has? Is it the product of intelligent design? Many participants in the dialogue are aware of the limitations of their field and do not claim to have all the answers. They hold that we can learn from each other. Some theologians are reformulating traditional ideas of God and human nature, taking the findings of science into account while trying to be faithful to the central message of their religious heritage.

In the Beginning: Why Did the Big Bang Occur?

Astronomers have presented convincing evidence that the early universe expanded very rapidly from a tiny, incredibly hot fireball fifteen billion years ago. But how can we explain the beginning itself, at which the laws of physics break down? The theist sees it as the moment of creation and the beginning of time. But the atheist can reply that there may have been an infinite span of time in which many universes have come into existence spontaneously, purely by chance. Alternatively, there may have been an oscillating universe in which a previous phase of contraction preceded the present expansive phase. In any case, the immense size and duration of the universe makes the brief existence of humanity on a minor planet seem insignificant. Religion meets science in the interpretation of cosmic history.

Quantum Physics: A Challenge to Our Assumptions about Reality?

Classical physics was deterministic and reductionistic in assuming that the behavior of all objects could be exactly predicted from accurate knowledge of their smallest components. Quantum physics, by contrast, acknowledges an inherent uncertainty in the prediction of events at the atomic and subatomic levels. It is also holistic in showing that the behavior of larger wholes is not simply the sum of the behavior of their parts, but involves distinctive system laws. Moreover, the quantum world can never be known as it is in itself, but only as it interacts with the observer in a particular experimental system. Quantum physics thus suggests the openness of the future, the interconnectedness of events, and the limitations of human knowledge. Some theistic interpreters propose that God determines the indeterminacies left open by the laws of quantum physics. Advocates of Eastern mysticism say that quantum holism supports their belief in the fundamental unity of all things. The new physics has led scientists, philosophers, and theologians to exciting discussions about time, causality, and the nature of reality.

Darwin and Genesis: Is Evolution God's Way of Creating?

We have all heard of debates between atheistic scientists and biblical literalists concerning Darwin's theory of evolution. But between these extremes have been many scientists and theologians who have believed in both evolution and God. From the scientific side, new concepts of complexity and self-organization have been used to portray the emergence of a hierarchy of higher levels. Some scientists have emphasized the role of information in molecular biology, evolutionary history, and embryonic development and suggest that the form of relationships is more important than the matter in which it is expressed. On the theological side, many authors have rejected the medieval view of a static universe in which all creatures were created in their present forms. They have explored the idea of a dynamic universe created over a long period of time by a God who is immanent in nature but also transcends nature. They hold that such a model of continuing creation is in harmony with the biblical understanding of the Holy Spirit as active in nature as well as in human life.

Human Nature: Are We Determined by Our Genes?

Comparisons of twins (having identical genes) with nontwin siblings (sharing half their genes) show that many forms of behavior are strongly influenced by genetic inheritance. In other studies, damage to particular areas of the brain and changes in the balance of chemicals in the brain have been shown to affect particular mental abilities dramatically. Some scientists conclude that human freedom is illusory. Others argue that although our decisions are severely constrained by our genes and our neurons, we can still make choices among a limited range of possibilities. The dependence of mental and spiritual life on biological processes calls into question the traditional dualisms of body/soul and matter/mind. Both scientists and theologians are elaborating views of the embodied, social self that are compatible with the findings of neuroscience and anthropology—and also compatible with the biblical view of the person as a unified activity of thinking, feeling, willing, and acting. The task here is to show that a human being can be at the same time a biological organism and a responsible self.

God and Nature: Can God Act in a Law-Abiding World?

A God who designed the universe and then let it go its own way is too remote from human life to be religiously significant. But can God have any continuing role in a world determined by scientific laws? One traditional answer is that God supports and works through lawful relationships to bring about predetermined goals. But is predestination consistent with human freedom and the presence of evil and suffering in the world? Some theologians have suggested that we must reject the classical idea of divine omnipotence. They speak of God's self-limitation in creating a world in which human freedom and lawful regularities are possible. Others have used concepts from current science to express ways in which God might act without violating the laws of science—for example, if God were the determiner of indeterminacies or the communicator of information. Some of the most creative work today involves collaboration between scientists and theologians in drawing from the ongoing experience of a religious community while taking seriously the discoveries of modern science. These crucial questions are explored in the chapters that follow.

Introduction

- *Forty-five percent of Americans believe that "God created man pretty much in his present form at one time within the last 10,000 years." Forty percent believe that "man developed over millions of years from less advanced forms of life but God guided the process." Ten percent believe that God had no part in the process. In other advanced industrial nations the fraction who take the Bible literally and reject evolution is far lower—only 7 percent in Great Britain, for example.[1]*

- *A 1997 survey of American scientists found that 39 percent believed in "a God to whom one may pray in expectation of receiving an answer"—as compared to 42 percent in a 1916 survey that used the same questions and an identical sampling procedure. The results challenged the widely held assumption that religious beliefs have fallen off drastically among scientists in the twentieth century.[2]*

- *In recent years there has been a growing literature exploring the relationships between science and religion. The average number of books published per year shown under the Library of Congress subject heading "Religion and Science" tripled from 71 during the 1950s to 211 in the 1990s.[3]*

In 1990, in the first chapter of *Religion in an Age of Science*, I proposed a fourfold typology as an aid to sorting out the great variety of ways in which people have related science and religion.[4] I kept the same classifications with only minor modifications in the revised and enlarged edition of the book in 1997.[5] In the present volume this typology is used as the organizing structure for every chapter.

1. *Conflict*. Biblical literalists believe that the theory of evolution conflicts with religious faith. Atheistic scientists claim that scientific evidence for evolution is incompatible with any form of theism. The two groups agree in asserting that a person cannot believe in both God and evolution, though they disagree as to which they will accept. For both of them, science and religion are enemies. These two opposing groups get most attention from the media, since a conflict makes a more exciting news story than the distinctions made by persons between these two extremes who accept both evolution and some form of theism.

2. *Independence*. An alternative view holds that science and religion are strangers who can coexist as long as they keep a safe distance from each other. According to this view, there should be no conflict because science and religion refer to differing domains of life or aspects of reality. Moreover, scientific and religious assertions are two kinds of language that do not compete because they serve completely different functions in human life. They answer contrasting questions. Science asks how things work and deals with objective facts; religion deals with values and ultimate meaning. Another version of the Independence thesis claims that the two kinds of inquiry offer complementary perspectives on the world that are not mutually exclusive. Conflict arises only when people ignore these distinctions—that is, when religious people make scientific claims, or when scientists go beyond their area of expertise to promote naturalistic philosophies. We can accept both science and religion if we keep them in separate watertight compartments of our lives. Compartmentalization avoids conflict, but at the price of preventing any constructive interaction.

3. *Dialogue*. One form of dialogue is a comparison of the methods of the two fields, which may show similarities even when

the differences are acknowledged. For example, conceptual models and analogies are used to imagine what cannot be directly observed (God or a subatomic particle, let us say). Alternatively, dialogue may arise when science raises at its boundaries limit-questions that it cannot itself answer (for example, Why is the universe orderly and intelligible?). A third form of dialogue occurs when concepts from science are used as analogies for talking about God's relation to the world. The communication of information is an important concept in many sciences; the pattern of unrepeatable events in cosmic history might be interpreted as including a communication of information from God. Or God can be conceived to be the determiner of the indeterminacies left open by quantum physics, without any violation of the laws of physics. Both scientists and theologians are engaged as dialogue partners in critical reflection on such topics, while respecting the integrity of each other's fields.

4. *Integration*. A more systematic and extensive kind of partnership between science and religion occurs among those who seek a closer integration of the two disciplines. The long tradition of natural theology has sought in nature a proof (or at least suggestive evidence) of the existence of God. Recently astronomers have argued that the physical constants in the early universe appear to be fine-tuned as if by design. If the expansion rate one second after the Big Bang had been ever so slightly smaller, the universe would have collapsed before the chemical elements needed for life could have formed; if the expansion rate had been even slightly higher, the evolution of life could not have occurred. Other authors start from a particular religious tradition and argue that some of its beliefs (ideas of divine omnipotence or original sin, for instance) should be reformulated in the light of science. Such an approach I call a *theology of nature* (within a religious tradition) rather than a *natural theology* (arguing from science alone). Alternatively, a philosophical system such as process philosophy can be used to interpret scientific and religious thought within a common conceptual framework. It will be evident that my own sympathies lie with Dialogue and

Integration (especially a theology of nature and a cautious use of process philosophy), but I hope that I have accurately described all four positions.

In 1995 John Haught offered a slightly different typology—one that may be easier to remember because all the terms start with the same letter.[6] His first two categories, Conflict and Contrast, are identical with those in my scheme. His third category, Contact, combines most of the themes in what I have called Dialogue and Integration. He introduces a fourth heading, Confirmation, by which he means not the confirmation of particular theological doctrines (as one might assume) but rather the vindication by science of background assumptions originally derived from theology—for example, belief in the rationality and intelligibility of the world, which I treat as a form of Dialogue.

Ted Peters proposes a more elaborate eightfold classification.[7] For example, he splits Conflict into three separate categories: Scientism, Scientific Creationism, and Ecclesiastical Authoritarianism. He also adds the category of Ethical Overlap, which is of course crucial in discussing applied science and technology. My typology was developed for fundamental science as a form of knowledge, not for applied science in its impact on society and nature. I explored ethical issues (especially environmental preservation and social justice) in my second set of Gifford Lectures, *Ethics in an Age of Technology*.[8] In the present book I will point out some connections between our understanding of nature and how we treat other creatures (environmental ethics), but the ethics of technology is not my concern here.

There is some advantage in using a larger number of classifications to allow greater discrimination. Willem Drees uses nine, arranged in three columns and three rows to emphasize experiential and cultural as well as cognitive interpretations of religion.[9] The disadvantage of introducing more categories is that the scheme becomes rather complicated, especially when it is used in examining a variety of scientific fields. Defining each category more narrowly yields greater precision, but one is more likely to find views that do not fit under any of them. Broader categories can include diverse cases more readily, but at the price of precision.

Some critics have suggested that the relations between science and religion are too complex and too context-dependent to be grouped under any classification scheme. They claim that the interactions vary too much among different historical periods and different scientific disciplines to show any general patterns. Wentzel van Huyssteen writes:

> The only way in which this complex but important relationship can really be adequately approached would be by looking at how it plays out contextually. This is also the reason why Ian Barbour's well-known, and helpful, fourfold taxonomy for relating religion and science through either conflict, dialogue, independence or integration may now be too generic, too universal, as categories that intend to catch the complexity of the ongoing exchange between these two dominant forces in our culture.[10]

Other authors maintain that scientific and religious ideas are only social constructions reflecting local cultural values, not objective descriptions of reality, so they cannot be related in any general or abstract way. Many "postmodern" writers hold that it is futile to seek knowledge or truth beyond the social constructions of particular cultures and historical periods.

The relations between science and religion are indeed complex, but I believe that examples of my four basic types can be found in each of the centuries since the rise of modern science and in each of the sciences. I hope that this typology will be helpful to readers new to this interdisciplinary field. A guidebook to any territory is not intended as a substitute for firsthand exploration, but is intended to help people find their way around. Guidebooks can be organized in a variety of ways, but they are necessarily selective and they may over-simplify the complexities of the real world. Each of the categories I have used groups together some very diverse proposals, among which the differences may be as significant as the similarities. We can acknowledge common patterns among various sciences and historical periods without ignoring the distinctiveness of each scientific discipline and historical context.

I will be focusing primarily on the Christian tradition, in which reflection on science has been far more extensive than in other traditions, both historically and today. It is indeed important to recognize the particularity of each religious tradition and to avoid the risk of superficial generalities in trying to include too much in a brief account. I believe that examples of each of the four categories can be found in the major world religions—especially in the monotheistic ones (Christianity, Judaism, and Islam), but also in Hinduism and Buddhism.[11] However, only a few such examples are included here. My attempt to categorize may itself reflect a Western bias. Authors from Eastern traditions might try to merge diverse viewpoints and to seek common ground among them.

The present volume is intended as an introduction to the field, briefer and more accessible than my earlier writings. At some points I have, with the publisher's permission, included revised passages from *Religion and Science* (1997), but the chapters here are organized in a new way, using the fourfold typology. In the first chapter, I set forth the basic types, with some historical examples and illustrations from recent authors. I then use the four types as the organizational framework in successive chapters on astronomy, quantum physics, evolutionary biology, and some of the human sciences (especially genetics and neuroscience). Each of those chapters starts by summarizing the scientific theories whose theological implications are explored. The final chapter looks at the more general problem of God's action in nature. The purpose of the book will be fulfilled if it encourages the reader's own reflection and further exploration of some of the issues and authors discussed on these pages.

Chapter One

Four Views of
Science and Religion

*T*his chapter describes four types of relationship between science and religion: Conflict, Independence, Dialogue, and Integration. Each type has several variants that differ significantly, but the variants have features in common that allow them to be grouped together. The applicability of this fourfold typology to particular scientific disciplines is explored in subsequent chapters.

Let me first describe two historical cases often cited as examples of Conflict. In both cases the historical record reveals a more complex relationship. The first is *the trial of Galileo* in 1633. Galileo advocated the new Copernican theory in which the earth and the planets revolve in orbits around the sun, rather than the accepted Ptolemaic theory in which the sun and planets revolve in orbits around the earth. One factor that contributed to the condemnation of Galileo was the authority of Aristotle, whose scientific writings, including those supporting Ptolemaic astronomy, had been greatly admired in Europe since the twelfth century. Another issue was the authority of scripture, especially the passages that implied that the earth is the center of the cosmos. But in the end the crucial factor was his challenge to the authority of the church.

In the centuries before Galileo a variety of views of scripture had been advanced. In the fourth century, Augustine (whom Galileo quoted) had said that when there appears to be a conflict between demonstrated knowledge and a literal reading of the Bible, scripture should be interpreted metaphorically. In commenting on the first chapter of Genesis, Augustine had said that the Holy Spirit was not concerned about "the form and shape of the heavens" and "did not wish to teach men things not relevant to their salvation." Medieval writers acknowledged diverse literary forms and levels of truth in scripture, and they offered symbolic or allegorical interpretations of many problematic passages. Galileo himself quoted a cardinal of his own day: "The intention of the Holy Ghost is to teach us how one goes to heaven, not how heaven goes." This aspect of Galileo's thought could be taken as an example of the Independence thesis, which distinguishes scientific from theological assertions. On astronomical questions, he said, the writers of the Bible had to "accommodate themselves to the capacity of the common people" by using "the common mode of speech" of their times. He held that we can learn from two sources, the Book of Nature and the Book of Scripture—both of which come from God and therefore cannot conflict with each other.

But Galileo introduced a qualification that opened the door to Conflict. He said that we should accept a literal interpretation of scripture unless a scientific theory that conflicts with it can be *irrefutably* demonstrated. He overstated the scientific certainty he could provide at a time when there was still considerable disagreement among astronomers. Moreover, the Catholic hierarchy felt under threat from the Protestant Reformation and was eager to reassert its authority. Some of the cardinals were sympathetic to Galileo's views, but the pope and several politically powerful cardinals were not. So he was finally condemned as much for disobeying the church as for questioning biblical literalism.[1]

The second case often cited as an example of Conflict is the debate over *Darwin's theory of evolution* in the nineteenth century. Some scientists and some religious leaders did indeed hold that evolution and religious beliefs are incompatible, but many in both groups did not. Three issues were at stake.[2]

1. *A Challenge to Biblical Literalism.* A long period of evolutionary change conflicts with the seven days of creation in Genesis. Some theologians of Darwin's day defended bibli-

cal inerrancy and rejected all forms of evolution, but they were in the minority. Most theological conservatives accepted symbolic rather than literal interpretations of these biblical passages and reluctantly accepted evolution, though they often insisted on the special creation of the human soul. The liberals, on the other hand, welcomed the advance of science and said that evolution was consistent with their optimistic view of historical progress. They were soon speaking of evolution as God's way of creating, which could be considered a version of what I have called Integration. They were also sympathetic to the work of biblical scholars who were studying evidence of the influence of the cultural and cosmological assumptions of the ancient Near East in the writings of biblical authors.

2. *A Challenge to Human Dignity.* In classical Christian thought, human beings were set apart from all other creatures, their unique status guaranteed by the immortality of the soul and the distinctiveness of human rationality and moral capacity. But in evolutionary theory humanity was treated as part of nature. No sharp line separated human and animal life, either in historical development or in present characteristics. Darwin and many of his successors stressed the similarities of human and animal behavior, though other biologists insisted on the distinctiveness of human language and culture. Copernican astronomy had demoted humanity from the center of the universe, and now Darwinian biology threatened human uniqueness in the order of nature. In Victorian England, many people saw the claim that we are "descended from apes" as a denial of the value of persons. "The survival of the fittest" seemed to undercut morality, especially when it was extrapolated into the social order to justify ruthless economic competition and colonialism.

3. *A Challenge to Design.* Within a static universe, the complex functioning of organisms and their harmonious adaptation to their surroundings offered a persuasive argument for an intelligent Designer. But Darwin showed that adaptation could be accounted for by an impersonal process of variation and natural selection. Darwin himself believed that God had designed the whole evolutionary process but not the detailed

structures of particular organisms. "I am inclined to look at everything as resulting from designed laws, with the details, whether good or bad, left to the working out of what we may call chance. . . . I cannot think that the world as we see it is the result of chance; yet I cannot look at each separate thing as the result of Design."[3] At the end of his life, Darwin seems to have been more agnostic, but through most of his life he accepted a reformulated version of the argument from design—an example of natural theology that I discuss under the heading Integration.

But some scientists went further and rejected even this broader concept of design. T. H. Huxley asserted that humanity is the product of impersonal and purposeless forces. The philosophy of *materialism* found new adherents among scientists in England and in Germany. The zoologist Ernst Haeckel held that matter and force are the only ultimate reality, and evolution provides an all-embracing explanation. "With this single argument the mystery of the universe is explained, the Deity annulled, and a new era of infinite knowledge ushered in."[4] In these varied responses to Darwin one can find examples of most of the ways of relating science and religion that are evident in the twentieth century.

CONFLICT

The Conflict thesis was promoted late in the nineteenth century by two influential books, J. W. Draper's *History of the Conflict between Religion and Science* and A. D. White's *A History of the Warfare of Science with Theology in Christendom*.[5] Recent historians point out that the evidence they cited was highly selective and that alternative views of the relation between science and religion were widely held during the centuries these authors describe.[6] Today the popular image of "the warfare of science and religion" is perpetuated by the media, for whom a controversy is more dramatic than the more subtle and discriminating positions between the extremes of scientific materialism and biblical literalism.

We could imagine a theological spectrum that would run as follows: naturalism (including materialism), pantheism, liberalism, neo-orthodoxy, traditionalism, conservatism, and biblical literalism (or fundamentalism). I have grouped the two extremes within the same category of Conflict—a pairing that may at first seem strange. I do this because *scientific materialism* and *biblical literalism* both claim that science and religion make rival literal statements about the same domain (the history of nature), so a person must choose between them. They agree in saying that a person cannot believe in both evolution and God. Each side gains adherents partly by its opposition to the other, and both use the rhetoric of warfare.

1. Scientific Materialism

Materialism is the assertion that matter is the fundamental reality in the universe. Materialism is a form of *metaphysics* (a set of claims concerning the most general characteristics and constituents of reality). *Scientific* materialism makes a second assertion: the scientific method is the only reliable path to knowledge. This is a form of *epistemology* (a set of claims concerning inquiry and the acquisition of knowledge). The two assertions are linked: if the only real entities are those with which science deals, then science is the only valid path to knowledge.

In addition, many forms of materialism express *reductionism*. Epistemological reductionism claims that the laws and theories of all the sciences are in principle reducible to the laws of physics and chemistry. Metaphysical reductionism claims that the component parts of any system determine its behavior. The materialist believes that all phenomena will eventually be explained in terms of the actions of material components, which are the only effective causes in the world. Analysis of the parts of any system has, of course, been immensely useful in science, but I will contend that the study of higher organizational levels in larger wholes is also essential.

Let us consider the assertion that the scientific method is the only reliable form of understanding. Science starts from reproducible public data. Theories are formulated as hypotheses that can be tested against experimental observations. Choices among theories are influenced by additional criteria of coherence, comprehensiveness, and fruitfulness in suggesting further research and applications. Religious beliefs are not acceptable, in this view, because religion lacks such public data, such experimental testing, and such criteria of evaluation. Science alone is objective, open-minded, universal, cumulative,

and progressive. Religious traditions, by contrast, are said to be subjective, closed-minded, parochial, uncritical, and resistant to change.
We will see that historians and philosophers of science have questioned this idealized portrayal of science, but many scientists accept it
and think that it undermines the credibility of religious beliefs.

Most of Carl Sagan's book (and television series) *Cosmos* is devoted
to a fascinating presentation of the discoveries of modern astronomy,
but at intervals Sagan interjects his own philosophical commentary.
He says that the universe is eternal or else its source is simply
unknowable. He attacks Christian ideas of God at a number of
points, arguing that mystical and authoritarian claims threaten the
ultimacy of the scientific method, which he says is "universally
applicable." Nature (which he capitalizes in the book) replaces God
as the object of reverence. Sagan expresses great awe at the beauty,
vastness, and interrelatedness of the cosmos.[7] In the television series
he sits at an instrument panel from which he shows us the wonders of
the universe. He is a new kind of high priest, not only revealing the
mysteries to us but telling us how we should live. We can be grateful
for Sagan's educational skills in bringing the findings of astronomy to
a wider public, and for his great ethical sensitivity and deep concern
for world peace and environmental preservation. But perhaps we
should question his unlimited confidence in the scientific method, on
which he says we should rely to bring in the age of peace and justice.

In Sagan's novel *Contact* and the 1997 film based on it, the scientific heroine detects radio signals from outer space and from them
she decodes the plans for a huge machine for space travel. The
machine is built, and with the help of spacetime changes in black
holes, she makes a brief trip to the center of the galaxy. The novel
and the film convey Sagan's sense of awe at the beauty and mystery of
the universe and his commitment to science and discovery. In the
novel a scientist who is asked if he has ever had a transforming religious experience replies that he did when he first understood gravita-
tion, relativity, and other theories, but "never apart from science."[8]
Institutional religion, by contrast, is represented by such dubious fig-
ures as fundamentalist protesters who think all space travel is the
work of the devil and a handsome TV evangelist who is more open to
science but has no formal education past grade school.

Much of Sagan's recent book, *A Demon-Haunted World: Science as a
Candle in the Dark*, is devoted to refuting pseudo-science, especially

astrology and claims of alien visitors and unidentified flying objects (UFOs). But many chapters are given to attacking religion, usually in its popular and superstitious forms. Sagan offers long accounts of belief in demons and witches in earlier centuries and faith healers and psychics today. But apart from one brief comment, he nowhere considers the writings of well-informed, university-based theologians who might be the intellectual counterparts of the scientists he admires.[9] Clearly Sagan sees science and religion as rivals, and he puts his faith and hope in the former.

As another example, consider the writing of the sociobiologist Edward O. Wilson. He traces the genetic and evolutionary origins of social behavior in insects, animals, and humans. He asks how self-sacrificial behavior could arise and persist among social insects such as ants if their individual reproductive future is thereby sacrificed. Wilson shows that such "altruistic" behavior enhances the survival of close relatives who share many of their genes (in an ant colony, for example). He believes that all human behavior can be reduced to and explained by its biological origins and present genetic structure. "It may not be too much to say that sociology and the other social sciences, as well as the humanities, are the last branches of biology" to be included in evolutionary theory.[10] The mind will be explained as "an epiphenomenon of the neural machinery of the brain."

Wilson holds that religious practices were a useful survival mechanism in humanity's earlier history because they contributed to group cohesion. But he says that the power of religion will be gone forever when religion is explained as a product of evolution; it will be replaced by "a philosophy of scientific materialism."[11] I would reply that if Wilson were consistent, he would have to say that the power of science will also be undermined when it is explained as a product of evolution. Do evolutionary origins really have anything to do with the legitimacy of either field? While Wilson has made an important contribution to our understanding of genetic constraints on human behavior, I suggest that he has elevated sociobiology into an all-encompassing explanation, leaving no room for the role of other facets of human life and experience. We will consider his views further in Chapters 5 and 6. Other examples of scientific materialism include the writings of Steven Weinberg (Chapter 2), Daniel Dennett (Chapters 4 and 5), Richard Dawkins (Chapters 4 and 6), Francis Crick (Chapter 5), and Peter Atkins (Chapter 6).

As I see it, these authors have failed to distinguish between *scientific* and *philosophical* questions. Scientists, in their popular writings, tend to invoke the authority of science for ideas that are not really part of science itself. Theism and materialism are alternative belief systems, each claiming to encompass all reality. In their epistemology, these authors assume that the scientific method is the only reliable source of knowledge—an assumption sometimes referred to by its critics as *scientism*. If science is the only acceptable form of understanding, explanation in terms of evolutionary history, biochemical mechanisms, or other scientific theories excludes all other forms of explanation. I suggest that the concept of God is not a hypothesis formulated to explain the relation between particular events in the world in competition with scientific hypotheses. Belief in God is primarily a commitment to a way of life in response to distinctive kinds of religious experience in communities formed by historic traditions; it is not a substitute for scientific research. Religious beliefs offer a wider framework of meaning in which particular events can be contextualized.

Every discipline is selective and has its limitations. Each abstracts from the totality of experience those features in which it is interested. The astronomer Arthur Eddington once told a delightful parable about a man studying deep-sea life using a net with a three-inch mesh. After bringing up repeated samples, the man concluded that there are no deep-sea fish smaller than three inches in length. Our methods of fishing, Eddington suggests, determine what we can catch. If science is selective, it cannot claim that its picture of reality is complete.[12]

In their metaphysics, these authors have extended scientific concepts beyond their scientific use to support comprehensive *materialistic philosophies*. The identification of the real with measurable properties that can be correlated by exact mathematical relationships started in the physical sciences but influenced scientists in other fields and continues today. But I would argue that these properties of matter have been abstracted from the real world by ignoring the particularity of events and the nonquantifiable aspects of human experience. We do not have to conclude that matter alone is real or that mind, purpose, and human love are only byproducts of matter in motion. Theism, in short, is not inherently in conflict with science, but it does conflict with a metaphysics of materialism.

2. Biblical Literalism

A very different kind of conflict arises from the literal interpretation of scripture. In the twentieth century, the Roman Catholic church and most of the mainline Protestant denominations have held that scripture is the human witness to the primary revelation that occurred in the lives of the prophets and the life and person of Christ. Many traditionalists and evangelicals insist on the centrality of Christ without insisting on the infallibility of the Bible. But fundamentalist churches and a large portion of some historic denominations in the United States (a majority in the case of the Southern Baptists) have maintained that scripture is inerrant throughout. The 1980s saw a growth of fundamentalist membership and political power. For many members of "the New Right" the Bible provides not only certainty in a time of rapid change but also a basis for the defense of traditional values in a time of moral disintegration (sexual permissiveness, prevalent drug use, high crime rates, and so forth). Biblical literalists say that evolutionary theory presupposes a philosophy of materialism and undermines belief in God's moral commandments.

In the Scopes trial in 1925, it was argued that the teaching of evolution in the schools should be forbidden because it is contrary to scripture. More recently, a new argument called *creation science* asserted that there is scientific evidence for the creation of the world within the last few thousand years. A law passed by the Arkansas legislature in 1981 required that "creationist theory" be given equal time with evolutionary theory in high school biology texts and classes. The law specified that creationism should be presented purely as scientific theory, with no reference to God or the Bible. In 1982 the U.S. District Court overturned the Arkansas law, primarily because it favored a particular religious view, violating the constitutional separation of church and state. Although the bill itself had made no explicit reference to the Bible, it used many phrases and ideas taken from Genesis. The writings of the leaders of the creationist movement had made clear their religious purposes.[13]

The court also ruled that creation science is *not legitimate science*. It concluded that the scientific community, not the legislature or the courts, should decide the status of scientific theories. It was shown that proponents of "creation science" had not even submitted papers to scientific journals, much less had them published. At the trial, scientific witnesses showed that a long evolutionary history is central in almost all fields of science, including astronomy, geology, paleontology, and biochemistry, as well as most branches of biology. They also replied to the purported scientific evidence cited by creationists. Claims of geological evidence for a universal flood and for the absence of fossils of transitional forms between species were shown to be dubious. In 1987 the U.S. Supreme Court struck down a Louisiana creationism law, saying that the law would have restricted academic freedom and supported a particular religious viewpoint.[14] In 1999 the Kansas school board ruled that no questions about evolution could be asked of students on state assessment tests. While it did not actually forbid the teaching of evolution, the ruling clearly hindered such teaching and is likely to be overturned by the courts.

I believe that creation science is a threat to both religious and scientific freedom. It is understandable that the search for certainty in a time of moral confusion and rapid cultural change has encouraged the growth of biblical literalism. Some of the same forces have contributed to the revival of Islamic fundamentalism and the enforcement of orthodoxy in Iran and elsewhere. We can see the danger to science when proponents of ideological positions try to use the power of the local school board or the state to reshape science. The scientific community can never be completely autonomous or isolated from its social context, yet it must be protected from political pressures that would dictate scientific conclusions. Science teachers must be free to draw from this larger scientific community in their teaching.[15]

The lawyer Phillip Johnson has raised valid criticisms of scientific materialists who promote atheistic philosophies as if they were part of science, but he goes on to attack evolutionary theory itself as inherently atheistic.[16] The biochemist Michael Behe has presented a detailed argument that complex interlocking systems (such as the long sequences of biochemical interactions in cells) could not have evolved by the natural selection of random variations and thus can only be the product of "intelligent design."[17] Their views are discussed in Chapter 4. I will suggest that though these authors are not biblical literalists they err in assuming that evolutionary theory and theism are incompatible. They, too, perpetuate the false dilemma of having to choose between science and religion. As Pope John Paul II has said, "Science can purify religion from error and superstition; religion can purify science from idolatry and false absolutes. Each can draw the other into a wider world, a world in which both can flourish."[18]

INDEPENDENCE

One way to avoid conflicts between science and religion is to keep the two fields in watertight compartments. They can be distinguished according to the *questions* they ask, the *domains* to which they refer, and the *methods* they employ. These are different kinds of distinction, but taken together they establish the independence and autonomy of the two fields. If there are two jurisdictions, each should tend to its own business and not meddle in the affairs of the other. Each mode of inquiry is selective and has its limitations. Compartmentalization is motivated not simply by the desire to avoid unnecessary conflicts but also by the desire to be faithful to the distinctive character of each area of life and thought. We will look first at science and religion as *separate domains* and then consider their *differing languages and functions.*

1. Separate Domains

Many *evangelicals* and *conservative Christians* today give scripture a central religious role without insisting on biblical literalism or defending creation science. They emphasize Christ's atoning death and the sudden conversion of the believer in accepting Christ as personal savior. They look to the transforming power of the gospel, which is neither threatened nor supported by modern science. For them, science and religion are totally independent aspects of human life.[19]

Protestant *neo-orthodoxy* has advocated a more explicit separation of science and religion, seeking to recover the Reformation emphasis on the centrality of Christ and the primacy of revelation while fully accepting the results of modern biblical scholarship and scientific research. According to Karl Barth and his followers, God can be known only as revealed in Christ and acknowledged in faith. God is the transcendent, the wholly other, unknowable except as self-disclosed. Religious faith depends entirely on divine initiative, not on human discovery of the kind occurring in science. The primary sphere of God's action is history, not nature. Scientists are free to carry out their work without interference from theology, and vice versa, since their methods and their subject matter are totally dissimilar. Science is based on human observation and reason, while theology is based on divine revelation.[20]

In this view, the Bible must be taken *seriously but not literally*. Scripture is not itself revelation; it is a fallible human record witnessing to revelatory events. The locus of divine activity was not the dictation of a text, but the lives of persons and communities (ancient Israel, the prophets, the person of Christ, and those in the early church who responded to him). The biblical writings reflect diverse interpretations of these events; we must acknowledge the human limitations of their authors and the cultural influences on their thought. Their opinions concerning scientific questions reflect the prescientific speculations of ancient times. We should read the opening chapters of Genesis as a symbolic portrayal of the basic relation of humanity and the world to God, a message about human creatureliness and the goodness of the natural order. These religious meanings can be separated from the ancient cosmology in which they were expressed.

Langdon Gilkey, in his earlier writing and in his testimony at the Arkansas creationism trial, expressed many of these themes. He made the following distinctions: (1) Science seeks to explain objective, public, repeatable data. Religion asks about the existence of order and beauty in the world and the experiences of our inner life (such as guilt, anxiety, and meaninglessness, on the one hand, and forgiveness, trust, and wholeness, on the other). (2) Science asks objective "how" questions. Religion asks personal "why" questions about meaning and purpose and about our ultimate origin and destiny. (3) The basis of authority in science is logical coherence and experimental adequacy. The final authority in religion is God and revelation, understood through persons to whom enlightenment and insight were given, and validated in our own experience. (4) Science makes quantitative predictions that can be tested experimentally. Religion must use symbolic and analogical language because God is transcendent.[21]

In the context of the Arkansas trial, it was an effective strategy to insist that science and religion ask different questions and use very different methods. That strategy provided methodological grounds for criticizing the attempts of biblical literalists to derive scientific conclusions from scripture. More specifically, Gilkey argued that the doctrine of creation is not a literal statement about the history of nature but a symbolic assertion that the world is good and orderly and dependent on God in every moment of time—a religious assertion essentially independent of both prescientific biblical cosmology and modern scientific cosmology. The views of neo-orthodoxy on creation are explored in "The Religious Meaning of Creation," in Chapter 2.

Another way of separating theological from scientific assertions is the distinction between *primary* and *secondary causality*, which is common in Catholic and neo-orthodox thought. God as primary cause is said to work through the secondary causes of the natural world that science studies. God is omnipotent and uses natural laws to achieve particular goals. Primary causality is on a totally different level from the interactions among entities in the world. This distinction will be explored in the context of evolution (Chapter 4) and in the theological writings of Gilson, Barth, and Farrer (Chapter 6). From the scientific side, a strong case that science and religion are separate domains has been presented by Stephen Jay Gould. His basic principle is non-overlapping magesteria (NOMA), where a magesterium is a domain of teaching authority (see Chapter 4).

2. Differing Languages and Functions

An alternative way of separating science and religion is to interpret them as languages that are unrelated because their functions are totally different. Among philosophers in the 1950s, the logical positivists took scientific statements as the norm for all cognitive assertions and dismissed as meaningless any statement not subject to empirical verification. The *linguistic analysts*, in response, insisted that differing types of language serve differing functions not reducible to each other. Each "language game" (as Ludwig Wittgenstein and his followers called it) is distinguished by the way it is used in a social context. Science and religion do totally different jobs, and neither should be judged by the standards of the other. *Scientific language* is used primarily for prediction and control. A theory is a useful tool for summarizing data, correlating regularities in observable phenomena, and producing technological applications. Science asks carefully delimited questions about natural phenomena. We must not expect it to do jobs for which it was not intended, such as providing an overall worldview, a philosophy of life, or a set of ethical norms. Scientists are no wiser than anyone else when they step out of their laboratories and speculate beyond strictly scientific work.[22]

The distinctive functions of *religious language*, according to the linguistic analysts, are to recommend a way of life, to elicit a set of attitudes, and to encourage allegiance to particular moral principles. Religious language arises from the ritual and practice of the worshiping community. It also expresses and leads to personal religious experience. One of the great strengths of the linguistic movement is that it does not concentrate on religious beliefs as abstract systems of thought but looks at the way religious language is actually used in the lives of individuals and communities. Linguistic analysts draw on empirical studies of religion by sociologists, anthropologists, and psychologists, as well as the literature produced within religious traditions.[23]

Scholars have looked at *the role of creation stories* in a variety of cultures. People in every age have sought to locate their lives within a cosmic order. Human interest in origins may be partly speculative or explanatory, but it is mainly motivated by the need to understand who we are in a larger framework of meaning and significance. Creation stories provide patterns for human behavior, archetypes of authentic human life. Religious communities appropriate and participate in such stories by symbolizing and enacting them in rituals. The exemplary patterns of primordial time are made present and celebrated in the liturgy. They serve a very different function in human life from those served by scientific research (see "The Function of Creation Stories," in Chapter 2).

Some scholars have studied diverse cultures and concluded that religious traditions are *ways of life* that are primarily practical and normative. Stories, rituals, and religious practices bind individuals in communities of shared memories, assumptions, and strategies for living. Other scholars claim that religion's primary aim is the transformation of the person. Religious literature speaks extensively of experiences of liberation from guilt through forgiveness, the replacement of anxiety by trust, and the transition from brokenness to wholeness.

Eastern traditions talk about liberation from bondage to suffering and self-centeredness in the experience of peace, unity, and enlightenment.[24] These are obviously activities and experiences having little to do with science.

A two-language approach also receives philosophical support from *instrumentalism*. According to instrumentalists, scientific theories are not representations of reality but useful intellectual and practical tools. They are convenient human constructs, calculating devices for correlating observations and making predictions. Models in science are not pictures of the world but useful fictions that can be discarded after they have been used to construct theories that will predict observations. Instrumentalists usually subscribe to pragmatism, in which the validity of a statement is judged by its usefulness in human life, not its correspondence to reality. An example of this view is developed in "Instrumentalist Views of Quantum Theory," in Chapter 3. When such an account of science is coupled with an instrumentalist or functionalist account of religion, the two fields are effectively separated from each other.

Finally, some authors refer to science and religion as *complementary perspectives*, invoking an analogy with complementarity in quantum physics. We will see in the next chapter that a subatomic entity such as an electron or a photon of light will sometimes behave like a particle and sometimes like a wave. It cannot be represented by one unified model or spatial specification; it can be represented only by a set of equations that predict the probability but not the exact value of a particular observation. Niels Bohr generalized his Complementarity Principle and applied it to alternative kinds of analysis of a single set of events, such as behavioristic and introspective models in psychology or divine justice and divine love in theology. We will find that some authors, such as the psychologists Malcolm Jeeves and Fraser Watts, have extended the idea further and speak of the complementarity of brain and mind (Chapter 5), and they defend the complementarity of scientific and religious views of the world (Chapter 6)—though they allow for limited forms of Dialogue that go beyond strict Independence.

I believe that the Independence thesis is a good starting point or first approximation. It preserves the distinctive character of each enterprise, and it is a useful strategy for responding to those who say conflict is inescapable. Religion does indeed have characteristic methods, questions, and functions that are distinct from those of science. But there are serious difficulties in the Independence proposals outlined above. As I see it, *neo-orthodoxy* rightly stresses the centrality of Christ and the prominence of scripture in the Christian tradition. It is more modest in its claims than biblical literalism, since it acknowledges the role of human interpretation in scripture and doctrine. But in most versions it holds that revelation and salvation occur only through Christ, which seems to me problematic in a pluralistic world. Most neo-orthodox authors emphasize divine transcendence and give short shrift to immanence. The gulf between God and the world is decisively bridged only in the Incarnation. While Karl Barth and his followers do indeed elaborate a doctrine of creation, their main concern is with the doctrine of redemption. Nature tends to be treated as the unredeemed setting for human redemption, though it may participate in the final fulfillment at the end of time.

The *two-language approach* can indeed help us to acknowledge the diversity of functions of religious language. Religion is a way of life and not simply a set of ideas and beliefs. But the religious practice of a community, including worship and ethics, presupposes distinctive beliefs. Against instrumentalism, which sees scientific theories and religious beliefs as human constructs useful for specific human purposes, I will in Chapter 3 advocate a critical realism that asserts that both communities make cognitive claims about realities beyond human life. We cannot remain content with science and religion as unrelated languages if they are languages about the same world. If we seek a coherent interpretation of all experience, we cannot avoid the search for a more unified worldview.

If science and religion were totally independent, the possibility of conflict would be avoided, but the possibility of constructive dialogue and mutual enrichment would also be ruled out. We do not experience life as neatly divided into separate compartments; we experience it in wholeness and interconnectedness before we develop particular disciplines to study different aspects of it. There are also biblical grounds for the conviction that God is Lord of our total lives and of nature, rather than of a separate "religious" sphere. The articulation of a theology of nature that encourages a strong environmental concern is also a critical task today. I will argue that none of the options considered above is adequate to that task.

DIALOGUE

Dialogue portrays more constructive relationships between science and religion than does either the Conflict or the Independence view, but it does not offer the degree of conceptual unity claimed by advocates of Integration. Dialogue may arise from considering the *presuppositions* of the scientific enterprise, or from exploring similarities between the *methods* of science and those of religion, or from analyzing *concepts* in one field that are analogous to those in the other. In comparing science and religion, Dialogue emphasizes similarities in presuppositions, methods, and concepts, whereas Independence emphasizes differences.

1. Presuppositions and Limit-Questions

Historians have wondered why modern science arose in the Judeo-Christian West among all world cultures. A good case can be made that *the doctrine of creation* helped to set the stage for scientific activity. Both Greek and biblical thought asserted that the world is orderly and intelligible. But the Greeks held that this order is necessary and that one therefore can deduce its structure from first principles. Only biblical thought held that God created both form and matter, meaning that the world did not have to be as it is and that the details of its order can be discovered only by observation. Moreover, while nature is real and good in the biblical view, it is not itself divine, as many ancient cultures held, and it is therefore permissible to experiment on it.[25]

We must be careful not to overstate the case for the role of Christian thought in the rise of science. Arab science made significant advances in the Middle Ages, while science in the West was often hampered by an otherworldly emphasis (although important practical technologies were developed, especially in some of the monastic orders). When modern science did develop in Europe, it was aided by the humanistic interests of the Renaissance, the growth of crafts, trade, commerce, and new patterns of leisure and education. Yet it does appear that the idea of creation gave a religious legitimacy to scientific inquiry. Many early scientists believed that in their work they were "thinking God's thoughts after him." Moreover, the Calvinist work ethic seems to have particularly supported science. In the Royal

Society, the earliest institution for the advancement of science, seven out of ten members were Puritans, and many were clergy.[26]

On the contemporary scene, *limit-questions* are raised by science but not answered within science itself. They are sometimes called "boundary questions," referring to methodological and conceptual boundaries as well as spatial and temporal boundaries. The Scottish theologian Thomas Torrance holds that science raises fundamental questions that it cannot answer. Science shows us an order that is both rational and contingent; its laws and initial conditions were not necessary. The combination of *contingency* and *intelligibility* prompts us to search for new and unexpected forms of rational order. Theologians hold that God is the creative ground of the contingent but rational order of the universe. "Correlation with that rationality in God goes far to account for the mysterious and baffling nature of the intelligibility inherent in the universe, and explains the profound sense of religious awe it calls forth from us and which, as Einstein insisted, is the mainspring of science."[27]

The Catholic theologian David Tracy sees a religious dimension in science. He holds that religious questions arise at the horizons or *limit-situations* of human experience. In everyday life, these limits are encountered in experiences of anxiety and confrontation with death, as well as in joy and basic trust. He describes two kinds of limit-situations in science: ethical issues in the uses of science and presuppositions or conditions for the possibility of scientific inquiry. Tracy maintains that the intelligibility of the world requires an ultimate rational ground. For the Christian, the sources for understanding that ground are the classic religious texts and the structures of human experience. All our theological formulations, however, are limited and historically conditioned.[28]

Theories concerning the Big Bang and the origin of the universe, in particular, raise questions related to temporal, spatial, and conceptual boundaries. Why is there a universe at all? Astronomy also raises questions about the intelligibility and contingency of the cosmos that are discussed in the next chapter.

2. Methodological and Conceptual Parallels

We have seen that exponents of scientific materialism typically hold that science is *objective*. Its theories are validated by clear-cut criteria and are tested by their agreement with indisputable, theory-free data. The data of science are unaffected by individual preferences or cultural influences. By contrast, religion seems to be highly *subjective* and strongly influenced by individual and cultural assumptions. Science is said to require detached observation and logical reasoning, whereas religion requires personal involvement in a particular tradition and set of practices.

But many historians, philosophers of science, and theologians have called into question this sharp contrast, arguing that science is not as objective nor religion as subjective as had been assumed. There are indeed differences of emphasis between the fields, but the distinctions are not absolute. Scientific data are theory-laden, not theory-free. Theoretical assumptions enter the selection, reporting, and interpretation of what are taken to be data. Moreover, theories do not arise from logical analysis of data but from acts of creative imagination in which *analogies and models* often play a role. Conceptual models help us to imagine what is not directly observable, especially in the realm of the very large (astronomy) and the very small (quantum physics).

Many of these same characteristics are present in religion. If the data of religion include religious experience, rituals, and scriptural texts, such data are even more heavily laden with conceptual interpretations. In religious language, too, *metaphors and models* are prominent, as discussed in my writing and in that of Sallie McFague and Janet Soskice.[29] Clearly, religious beliefs are not amenable to strict empirical testing, but they can be approached with some of the same spirit of inquiry found in science. The scientific criteria of coherence, comprehensiveness, and fruitfulness have their parallels in religious thought.

Thomas Kuhn's influential book, *The Structure of Scientific Revolutions*, maintained that both theories and data in science are dependent on the prevailing paradigms of the scientific community. He defined a *paradigm* as a cluster of conceptual, metaphysical, and methodological presuppositions embodied in a tradition of scientific work. With a new paradigm, the old data are reinterpreted and seen in new ways, and new kinds of data are sought. In the choice between paradigms, there are no rules or determinative criteria. Their evaluation is an act of judgment by the scientific community. An established paradigm is resistant to falsification, since discrepancies between theory and data can be set aside as anomalies or reconciled by introducing ad hoc hypotheses.[30] I have suggested that religious traditions can also be looked on as communities that share a common paradigm. The interpretation of the data (such as religious experience and historical events) is even more paradigm-dependent in religion than in science. Ad hoc assumptions are often introduced to reconcile apparent anomalies, so religious paradigms are even more resistant to falsification, but they are not totally immune to challenge.[31]

Several authors have invoked such *methodological parallels* between science and religion. The physicist and theologian John Polkinghorne gives examples of personal judgment and theory-laden data in both fields. The data for a religious community are its scriptural records and its history of religious experience. Similarities exist between the fields in that "each is corrigible, having to relate theory to experience, and each is essentially concerned with entities whose unpicturable reality is more subtle than that of naive objectivity."[32] The philosopher Holmes Rolston holds that religious beliefs interpret and correlate experience, much as scientific theories interpret and correlate experimental data. Beliefs can be tested by criteria of consistency and congruence with experience. But Rolston acknowledges that personal involvement is more total in the case of religion, since the primary goal is the reformation of the person. Moreover, there are other significant differences: science is interested in causes, for example, while religion is interested in personal meanings.[33]

The *status of the observer* in science has also been reconsidered. The earlier accounts had identified objectivity with the separability of the observer from the object of observation. But in quantum physics the influence of the process of observation on the system observed is crucial. In relativity, the most basic measurements, such as the mass, velocity, and length of an object, depend on the frame of reference of the observer. Stephen Toulmin traces the change from the assumption of a detached spectator to the recognition of the participation of the observer; he cites examples from quantum physics, ecology, and the social sciences. Every experiment is an action in which we are agents, not just observers. The observer is a participant inseparable from the object of observation.[34] Examples from quantum physics concerning the role of the observer are given in Chapter 3.

All these authors acknowledge that there are *differences in the methods* of science and religion. Science is clearly more objective than religion in each of the senses mentioned above. The kinds of data from which religion draws are radically different from those in science, and the possibility of testing religious beliefs is much more limited. Religion is more than an intellectual system, because its goal is personal transformation and a way of life. But all these authors insist that there are also *significant parallels* in the methods of the two fields, including the use of criteria of consistency and congruence with experience. They hold that theology at its best is a self-critical, reflective enterprise that can be open to new insights, including those originating in the sciences.

In addition to methodological parallels, conceptual parallels between science and religion have been explored by many recent authors. The *communication of information* in science, for example, offers some interesting parallels with the biblical view of the divine Word in creation. Information is an important concept in several fields of science (DNA in organisms, programs in computers, the neural structures in the brain), as described in Chapter 4 (see "The Concept of Information"). Polkinghorne has suggested that God's activity in the world can be thought of as the communication of information (see "God as Communicator of Information," in Chapter 6). The *self-organization* of complex systems (especially nonlinear systems far from equilibrium), described in Chapter 4, has been correlated with a model of God as the designer of a self-organizing process (Chapter 6). In short, methodological and conceptual parallels, like presuppositions and limit-questions, offer the possibility of significant dialogue between science and religion while preserving the integrity of each field.

INTEGRATION

Some authors call for reformulations of traditional theological ideas that are more extensive and systematic than those envisaged by advocates of Dialogue discussed above. There are three distinct versions of Integration. In *natural theology*, it is claimed that the existence of God can be inferred from (or is supported by) the evidence of design in nature, of which science has made us more aware. In a *theology of nature*, the main sources of theology lie outside science, but scientific theories may strongly affect the reformulation of certain doctrines, particularly the doctrines of creation and human nature. In a *systematic synthesis*, both science and religion contribute to the development of an inclusive metaphysics, such as that of process philosophy.

I. Natural Theology

There are many examples of natural theology from previous centuries. Thomas Aquinas held that some of God's characteristics can be known only from revelation in scripture but that the existence of God can be known by reason alone. One form of his *cosmological argument* asserted that every event must have a cause, so we must posit a First Cause if we are to avoid infinite regress. Another form said that the whole chain of natural causes (finite or infinite) is contingent and might not have been; it is dependent on a being that exists necessarily. Aquinas's *teleological argument* (from *telos*, Greek for *purpose* or *goal*) starts from orderliness and intelligibility as general characteristics of nature but goes on to cite specific evidence of design in nature.[35]

The founders of modern science frequently expressed admiration for the harmonious coordination of nature, which they saw as God's handiwork. Newton said that the eye could not have been contrived without skill in optics, and Robert Boyle extolled the evidences of benevolent design throughout the natural order. By the eighteenth century, the world was viewed as a clockwork mechanism, with the clockmaker God of deism as its designer. But the philosopher David Hume offered an extended critique of the argument from design. He observed that the organizing principle responsible for patterns in nature might be *within* organisms, not *external* to them. At most, he said, the argument would point to the existence of a finite god or many gods, not the omnipotent Creator of monotheism. He also claimed that the presence of evil and suffering in the world undermines the traditional design argument.[36]

Despite Hume's critique, *the argument from design* remained very popular. William Paley said that if one finds a watch on a heath, one is justified in concluding that it was designed by an intelligent being. Similarly, he said, when one considers the human eye, its many complex parts coordinated to the single purpose of vision, one can only conclude that it is the product of intelligent design. Paley cited many other examples of the coordination of structures fulfilling functions useful to living organisms.[37] It was Darwin, of course, who dealt the most serious blow to the argument, for he showed that adaptation can be explained by random variation and natural selection. But Darwin himself defended (at least until his last few years) a reformulated version of the argument. As noted earlier, he said that God did not design the particular details of individual species but designed the laws of the evolutionary processes through which the species were formed, leaving the details to chance.

Among contemporary philosophers, Richard Swinburne has given an extended defense of natural theology. He starts by discussing *confirmation theory* in the philosophy of science. In the development of science, new evidence does not make a theory certain. Instead, a theory has an initial plausibility, and the probability that it is true increases or decreases as additional evidence is acquired (Bayes's theorem). Swinburne suggests that the existence of God has an initial plausibility because of its simplicity and because it gives a coherent explanation of the world in terms of the intentions of an agent. He then argues that the evidence of order in the world increases the probability of the theistic hypothesis. He also maintains that science cannot account for the presence of conscious beings in the world. "Something outside the web of physical laws" is needed to explain the rise of consciousness. Finally, religious experience provides "additional crucial evidence." Swinburne concludes, "On our total evidence, theism is more probable than not."[38]

The most recent rendition of the design argument is the *Anthropic Principle* in cosmology. Astrophysicists have found that life in the universe would have been impossible if some of the physical constants and other conditions in the early universe had differed even slightly in the values they had. The universe seems to be "fine-tuned" for the possibility of life. For example, Stephen Hawking says that if the rate of expansion one second after the Big Bang had been smaller by even one part in a hundred thousand million million, the universe would have recollapsed before life could have formed.[39] The physicist Freeman Dyson draws the following conclusion from such findings:

I conclude from the existence of these accidents of physics and astronomy that the universe is an unexpectedly hospitable place for living creatures to make their home in. Being a scientist, trained in the habits of thought and language of the twentieth century rather than the eighteenth, I do not claim that the architecture of the universe proves the existence of God. I claim only that the architecture of the universe is consistent with the hypothesis that mind plays an essential role in its functioning.[40]

John Barrow and Frank Tipler present many other cases in which there were *extremely critical values* of various forces in the early universe.[41] The philosopher John Leslie defends the Anthropic Principle as a design argument. But he points out that an alternative explanation would be the assumption of many worlds (either in successive cycles of an oscillating universe or in separate universes existing simultaneously). These worlds might differ from each other, and we just happen to be in one that had the right variables for the emergence of life. Moreover, some of these apparently arbitrary conditions may be necessitated by a more basic unified theory that has yet to be discovered.[42] The Anthropic Principle and its critics are discussed in detail in Chapter 2. In Chapter 4 we will consider the concept of "evolutionary design." Other versions of natural theology are explored in Chapter 6.

Natural theology has a great appeal in a world of religious pluralism, since it starts from scientific data on which we might expect agreement despite cultural and religious differences. Moreover, it is consistent with the personal response of awe and wonder that many scientists experience in their work. Proponents of the design argument today do not claim that it offers conclusive evidence for theism; they assert the more modest claim that belief in a Designer is more plausible than (or at least as plausible as) alternative interpretive proposals. This can help to overcome obstacles to belief and can lead to a greater openness to other forms of religious experience and to participation in a religious community. On the other hand, the limitations of the argument should be acknowledged. Taken alone, it can at best lead only to the God of deism, the intelligent Designer remote from the world. However, it can be combined with theistic beliefs based on personal religious experience and a historical tradition. Exponents of a theology of nature can make use of design arguments but are not likely to give them a central place in their life and thought.

2. Theology of Nature

A theology of nature does not start from science, as natural theology usually does today. Instead, it starts from a religious tradition based on religious experience and historical revelation. But it holds that some traditional doctrines need to be *reformulated* in the light of current science. Here science and religion are considered to be relatively independent sources of ideas, but with some areas of overlap in their concerns. In particular, the doctrines of creation and human nature are affected by the findings of science. If religious beliefs are to be in harmony with scientific knowledge, more extensive adjustments or modifications are called for than those introduced by proponents of the Dialogue thesis. It is said that the theologian should draw from broad features of science that are widely accepted, rather than risk adapting to limited or speculative theories that are more likely to be abandoned in the future. Theological doctrines must be consistent with the scientific evidence even if they are not directly implied by current scientific theories.

Our understanding of the general characteristics of nature affects our models of God's relation to nature. Nature is today understood to be *a dynamic evolutionary process* with a long history of emergent novelty, characterized throughout by chance and law. The natural order is ecological, interdependent, and multileveled. These characteristics modify our representation of the relation of both God and humanity to nonhuman nature. This in turn affects our attitudes toward nature and has practical implications for environmental ethics. The problem of evil is also viewed differently in an evolutionary rather than a static world.

For the biochemist and theologian Arthur Peacocke, the starting point of theological reflection is past and present religious experience in an ongoing religious community. Religious beliefs are tested by community consensus and by criteria of coherence, comprehensiveness, and fruitfulness. But Peacocke is willing to reformulate traditional beliefs in response to current science. He discusses at length how *chance* and *law* work together in cosmology, quantum physics, nonequilibrium thermodynamics, and biological evolution. He describes the emergence of distinctive forms of activity at higher levels of complexity in the multilayered hierarchy of organic life and mind.

He gives chance a positive role in the exploration and expression of potentialities at all levels. God creates through the whole process of law and chance, not by intervening in gaps in the process. God creates in and through the processes of the natural world that science unveils.

Peacocke speaks of chance as God's radar sweeping through the range of possibilities and evoking the diverse potentialities of natural systems. In other images, artistic creativity is used as an analogy for unpredictable purposefulness.[43] Peacocke's theological ideas are explored in Chapter 4, which also presents the concept of *top-down causality* in organisms. Chapter 6 discusses the extension of these ideas in the model of "God as Top-Down Cause," proposed by Peacocke, Philip Clayton, and others.

Another theological proposal starts from the analysis of *indeterminacy* in quantum theory. I argue in Chapter 3 that the uncertainty of predictions in the world of subatomic particles represents a genuine indeterminacy in nature, and not merely a limitation in human knowledge of events in nature that are themselves strictly determined. The last section of Chapter 3 describes the claim that God is the ultimate determiner of the indeterminacies at the quantum level—without violating the laws of nature, which specify only probabilities within a range of values rather than exact values. The model of "God as Determiner of Indeterminacies" is developed further in Chapter 6, drawing on the writing of Robert Russell, George Ellis, Nancey Murphy, and Tom Tracy. Such direct uses of scientific ideas in theology go beyond the comparison of parallel concepts discussed under Dialogue and belong under the rubric of Integration. Because they do not offer an argument from scientific evidence to the existence of God, they exemplify a theology of nature rather than a natural theology.

Other recent versions of a theology of nature can be found among feminist authors. They have pointed to correlations among the dualisms that have been so pervasive in Western thought: mind/body, reason/emotion, objectivity/subjectivity, domination/submission, power/love. In each case, the first term of each pair (*mind, reason, objectivity, domination, power*) is identified in our culture as male, the second term (*body, emotion, subjectivity, submission, love*) as female. A historically patriarchal culture, in which men have held most of the positions of power, has perpetuated a predominantly male image of

God. Moreover, the first term of each pair is thought to be character-istic of science, especially in its attempt to dominate and control nature. Many feminists hold that the exploitation of women and the exploitation of nature have common ideological roots in the West. The radical ecofeminists turn to indigenous cultures for feminine symbols of the divine and for recovery of the sacred in nature.[44] Reformist feminists, on the other hand, believe that the patriarchal features of historical Christianity can be rejected without rejecting the whole tradition. I am particularly indebted to reformist feminists such as Sallie McFague and Rosemary Radford Ruether.[45]

I believe that a theology of nature must draw from both science and religion in the task of formulating an *environmental ethics* relevant to today's world. Only science can supply the data for evaluating threats to the environment arising from our technologies and our lifestyles, but religious beliefs can significantly affect our attitudes toward nature and our motivation for action. Environmentalists have rightly criticized classical Christianity for drawing too sharp a line between human and nonhuman nature and for emphasizing God's transcendence at the expense of immanence. New religious interpre-tations of human nature that take into account the evolutionary con-tinuity and ecological interdependence of human and nonhuman life are set forth in Chapter 5. New ways of representing both divine transcendence and immanence in the context of current science are investigated in Chapter 6.

The idea of human dominion over nature in Genesis 1:28 has sometimes been used to justify unlimited domination in which other creatures are treated simply as means to human ends. How-ever, many recent authors have urged the recovery of biblical themes that give strong support to environmentalism.[46] *Stewardship of nature* is called for because the earth belongs ultimately to the God who created it. We are only trustees and stewards, responsible for its welfare and accountable for our treatment of it. The Sabbath is a day of rest for all living things, not just for humanity. *Celebration of nature* goes beyond stewardship, asserting that nature is valuable in itself, not simply for human uses. The first chapter of Genesis ends with an affirmation of the goodness of the created order encompassing all forms of life, and many of the psalms celebrate the rich diversity of the natural world, including strange creatures that are of no use to us.

Sacramental views of nature attribute even greater value to the natural world, affirming that the sacred is present in and under it. Eastern Orthodoxy, Celtic Christianity, and some Anglican authors hold that all of nature, and not just the bread, wine, and water of the sacraments, can be a vehicle of God's grace. The same God encountered in the life of Christ and the church can be encountered in nature. These varied themes in an environmental ethics for today are mentioned at several points in this volume, but I have explored them more extensively elsewhere.[47]

3. Systematic Synthesis

A more systematic integration can occur if both science and religion contribute to a coherent worldview elaborated in a comprehensive metaphysics. Metaphysics is the search for a set of general concepts in terms of which diverse aspects of reality can be interpreted. An inclusive conceptual scheme is sought that can represent the fundamental characteristics of all events. Metaphysics as such is the province of the philosopher rather than of either the scientist or the theologian, but it can serve as an arena of common reflection.

In the thirteenth century, Thomas Aquinas articulated an impressive metaphysics that has remained influential in Catholic thought. His voluminous writings systematically integrated ideas from earlier Christian authors with the best philosophy and science of his day, derived largely from the works of Aristotle, which had been rediscovered in the West through Arabic translations.[48] But I would argue that Aquinas's thought expressed dualisms of matter/spirit, body/soul, temporality/eternity, and nature/humanity that have been only partially overcome in more recent Thomistic thought (see Chapter 5).

Process philosophy is a promising candidate for a mediating role today because it was itself formulated under the influence of both scientific and religious ideas. Alfred North Whitehead was familiar with quantum physics and its portrayal of reality as a series of momentary events and interpenetrating fields rather than separate particles. In his thought, processes of change and relationships between events are more fundamental than enduring self-contained objects. For him, as for evolutionary thinkers, nature is a dynamic web of interconnected events, characterized by novelty as well as order. Whitehead and his followers are critical of reductionism; they defend organismic categories applicable to activities at all levels of organization.[49]

Process thought holds that the basic constituents of reality are not two kinds of enduring entity (mind/matter dualism) or one kind of enduring entity (materialism), but *one kind of event* with *two aspects or phases*. This philosophy is monistic in portraying the common character of all events, but it recognizes that these events can be organized in diverse ways, leading to an organizational pluralism of many levels. All integrated entities, such as organisms (but not unintegrated aggregates, such as stones) have an inner and an outer reality, but these take very different forms at different levels. Interiority varies widely, starting from rudimentary memory, sentience, responsiveness, and anticipation in simple organisms and leading to consciousness, which requires a central nervous system. Viewed from within, interiority can be construed as a moment of experience, though *conscious* experience occurs only at high levels of organization, and reflective self-consciousness is unique to humanity. Genuinely new phenomena emerge in evolutionary history, but the basic metaphysical categories apply to all events.[50]

For process thinkers, God is *the source of novelty and order*. Creation is a long and incomplete process. God elicits the self-creation of individual entities, thereby allowing for freedom and novelty as well as order and structure. God is not the transcendent Sovereign of classical Christianity. God interacts reciprocally with the world, an influence on all events though never the sole cause of any event. Process metaphysics understands every new event to be jointly the product of the entity's past, its own action, and the action of God. Here God transcends the world but is immanent in the world in a specific way in the structure of each event. We do not have a succession of purely natural events, interrupted by gaps in which God alone operates.

Charles Hartshorne has elaborated a version of process thought with a "dipolar" concept of God: unchanging in purpose and character, but changing in experience and relationship.[51] Process thinkers reject the idea of divine omnipotence; they believe in a God of persuasion rather than coercion, and they have provided distinctive analyses of the place of chance, human freedom, evil, and suffering in the world. Christian process theologians such as John Cobb and David Griffin point out that the power of love, as exemplified in the cross, is precisely its ability to evoke a response while respecting the integrity of other beings. They also hold that divine immutability is not a characteristic of the biblical God who is intimately involved with history.[52]

Process thought can also make a distinctive contribution to *environmental ethics*. Human life and nonhuman life are not separated by any absolute line. If other creatures are centers of experience, they too are of intrinsic value and not just of instrumental value to humanity. Another process theme with environmental implications is the emphasis on interdependence. Whereas traditional theology has emphasized transcendence (without ignoring immanence), process thought leads to an emphasis on divine immanence in nature (without ignoring transcendence), which would encourage greater respect for nature. Various aspects of process philosophy and process theology are developed in the concluding sections of Chapters 4, 5, and 6.

Having surveyed a wide range of positions and issues in this chapter, I conclude by summarizing my own position concerning each of the four types described:

1. *Conflict*. I have argued that both scientific materialists and biblical literalists have failed to recognize significant distinctions between scientific and religious assertions. The scientific materialists have promoted a particular philosophical commitment as if it were a scientific conclusion, and the biblical literalists have promoted a prescientific cosmology as if it were an essential part of religious faith.

2. *Independence*. Neo-orthodoxy rightly says that in the Christian community it is in response to the person of Christ that our lives can be changed. As I see it, the center of the Christian life is an experience of reorientation, the healing of our brokenness in new wholeness, and the expression of a new relationship to God and neighbor. But the centrality of redemption does not need to lead us to ignore creation, for our personal and social lives are intimately bound to the rest of the created order. Nature is more than the impersonal stage for the drama of personal redemption. We are part of a drama that includes all creatures. Linguistic analysts rightly point to the distinctive functions of story, ritual, and practice in the life of a religious community. A reli-

gious tradition is indeed a way of life and not a set of abstract ideas. But a way of life presupposes beliefs about the nature of reality and cannot be sustained if those beliefs are no longer credible.

3. *Dialogue.* The presuppositions of science and the limit-questions it raises (such as questions regarding the contingency and intelligibility of the universe) were important historically and still are, but these issues are foundational and may seem rather abstract to most people today. I consider methodological parallels more significant because they affect our understanding of scientific and theological inquiry and thus our view of the relation between the fields. Conceptual parallels between particular scientific theories and particular theological beliefs are even closer to the daily work of scientists and theologians, and they are prominent in some of the most creative interactions between the disciplines today.

4. *Integration.* As a form of natural theology, current arguments from design do not claim to offer a proof of the existence of God, but they suggest that theism is as plausible as (or more plausible than) other interpretations of the pattern of cosmic history. This may help to answer the claims of philosophical materialism, but it leaves out the most central aspects of personal religious experience in a religious tradition. A theology of nature seems to me more promising because it starts from the life of a religious community and asks how its beliefs may need to be reformulated in the context of modern science. Contemporary ideas of evolutionary history, law and chance, and the many-leveled character of biological organisms are relevant to doctrines of creation, providence, and human nature, as well as to environmental ethics.

As we attempt to articulate a theology of nature, a systematic metaphysics such as process philosophy can help us in the search for a coherent vision. But neither science nor religion should be equated with a metaphysical system. There are dangers if either scientific or religious ideas are distorted to fit a preconceived synthesis that claims to encompass all reality. We must always keep in mind the rich diversity of our experience. We distort it if we cut it up into separate realms or watertight compartments, but we also distort it if we force it into a neat intellectual system. A coherent vision of reality must allow for the distinctiveness of differing types of experience.

In examining particular sciences in each of the chapters that follow, I will indicate my reasons for disagreeing with the Conflict thesis. I will point out what I consider to be valid themes in the Independence position, even though I do not accept its conclusions. I will describe some significant proposals for Dialogue, especially those suggesting methodological and conceptual parallels. Finally, I will draw from advocates of Integration in the reformulation of doctrines of creation, human nature, and (more briefly) environmental ethics, including a cautious use of ideas from process philosophy.

Chapter Two

Astronomy and Creation

$\mathcal{M}$ ost astronomers from Ptolemy, Copernicus, and Galileo through the eighteenth century assumed that the universe was relatively small in size and young in age. In the nineteenth century, speculative theories of a larger and older universe were proposed. In the twentieth century, evidence of the immense size and age of the universe has accumulated, and new cosmological theories have raised significant issues in relation to religious beliefs.

In 1917, Willem de Sitter, working with Einstein's general relativity equations, found a solution that predicted an expanding universe. In 1929, Edwin Hubble, examining the "red shift" of light from distant nebulae, formulated Hubble's law: the velocity of recession of a nebula is proportional to its distance from us. Extrapolating backward in time, scientists have concluded that the universe seems to be expanding from a common origin about fifteen billion years ago that has come to be known as the Big Bang. In 1965, Arno Penzias and Robert Wilson discovered a faint background of microwaves coming from all directions in space. The spectrum of those waves corresponded very closely to the residual radiation that had been predicted from relativity theory. The radiation is the cosmic fireball's afterglow, cooled by its subsequent expansion.

Indirect evidence concerning the very early moments of the Big Bang have come from both theoretical and experimental work in high-energy physics. This research shows that there are *four basic physical forces:* (1) the electromagnetic force responsible for light and the behavior of charged particles; (2) the weak nuclear force responsible for radioactive decay; (3) the strong nuclear force that binds protons and neutrons into nuclei; and (4) the gravitational force evident in the long-distance attraction between masses. In 1967, Steven Weinberg and Abdus Salam showed that the electromagnetic and weak forces could be unified within an *electroweak theory.* The theory predicted the existence of two massive particles, dubbed W and Z, which mediate between the two kinds of force. In 1983, Carlo Rubbia and coworkers found evidence for W particles among the products of high-energy collisions in a particle accelerator in Geneva.[1]

There has been some progress in attempts to unify the electroweak and strong forces in a *Grand Unified Theory.* The unification would be mediated by very massive X particles, which could exist only at energies higher than those in any existing accelerator. Bringing gravity and the other three forces together within one *supersymmetry theory* (sometimes called a Theory of Everything) has been more difficult. One candidate is *superstring theory,* which postulates incredibly heavy, tiny, one-dimensional strings that can split or loop. The unification of quantum theory and gravity requires a framework of ten dimensions, but superstring theory suggests that six of these dimensions immediately contracted to negligible size, leaving only our present four dimensions of space and time. There is no direct experimental evidence for strings; the energy required for their existence would be far beyond that available in laboratory experiments, but it would have been present at the very earliest instants of the Big Bang. Physicists have a strong commitment to simplicity, unity, and symmetry, which motivates the search for a unified theory even when direct experimentation is impossible.

Putting together the evidence from astronomy and high-energy physics, a plausible *reconstruction of cosmic history* can be made for events starting three minutes after the Big Bang, when protons and neutrons were combining to form nuclei. Five hundred thousand years later, atoms were coming into existence. One billion years from the beginning, galaxies and stars were being formed, and then planets at ten billion years. After another two billion years, microscopic forms of life were beginning to appear on our planet, and biological evolution was under way.

The farther back we go before three minutes, the more tentative are the theories, because they deal with states of matter and energy increasingly farther from anything we can duplicate in the laboratory. Protons and neutrons were probably forming from their constituent quarks at 10^{-4} seconds (a ten-thousandth of a second from the beginning), when the temperature had cooled to 10^{12} (a thousand billion) degrees. This fantastically dense sea of hot quarks would have been formed at about 10^{-10} seconds from an even smaller and hotter fireball. The inflationary theories proposed by Alan Guth and Andrei Linde suggest that the universe underwent a very rapid expansion at about 10^{-35} seconds due to the tremendous energy released in the breaking of symmetry when the strong force separated from the other forces.[2] Before 10^{-35} seconds, the temperature would have been so high that all the forces except gravity were of comparable strength. This is the period to which a Grand Unified Theory would apply. We have almost no idea of events before 10^{-43} seconds, when the temperature was 10^{32} degrees. The whole universe was the size of an atom today, and the density was an incredible 10^{96} times that of water.[3]

What happened at $t = 0$? In standard Big Bang theory, $t = 0$ is a *singularity* to which the laws of physics do not apply. It would have been a dimensionless point of pure radiation of infinite density. Some theologians see common ground with cosmologists (astronomers and physicists interested in the early history of the cosmos) in the idea that the universe had a beginning. Pope Pius XII said the Big Bang theory supports the biblical idea of creation.[4] The astrophysicist Robert Jastrow has argued that "the astronomical evidence leads to a biblical view of the origin of the world." He ends his book *God and the Astronomers* with this striking passage:

> At this moment it seems as though science will never be able to raise the curtain on the mystery of creation. For the scientist who has lived by his faith in the power of reason, the story ends like a bad dream. He has scaled the mountains of ignorance; he is about to conquer the highest peak; as he pulls himself over the final rock, he is greeted by a band of theologians who have been sitting there for centuries.[5]

But other people—both scientists and theologians—are uncomfortable when the doctrine of creation is identified with a particular scientific theory. In this chapter we will examine how adherents of each of our four basic views of science and religion have responded to recent cosmological theories.

CONFLICT

One form of conflict between science and religion is presented by atheists who say that it was entirely a matter of chance that the balance of forces in the early universe produced conditions favorable to the appearance of life and intelligence. At the other extreme, defenders of biblical inerrancy have claimed that relativity theory makes it possible to harmonize the Genesis account with the Big Bang.

I. A Universe by Chance

Forty years ago, some cosmologists thought that they could avoid a moment of creation by postulating an infinite span of time. Their *steady-state theory* proposed that hydrogen atoms come into being, slowly and continuously, throughout an infinite time and space. Fred Hoyle, in particular, defended the theory long after most of his colleagues had abandoned it.[6] Hoyle's writings make clear that he favored the steady-state theory not just on scientific grounds, but partly because he thought infinite time was more compatible with his own atheistic beliefs. Today variants of the Big Bang theory have clearly won the day.

It is a remarkable fact of our universe that its physical constants are precisely right for the emergence of life and intelligence. If the strong nuclear force or the expansion rate, for example, had been even very slightly larger or smaller, organic life would have been impossible. We will later examine the claim that these "fine-tuned phenomena" offer a new argument for design. But some cosmologists suggest that there may have been *many universes* with differing parameters, and we just happen by chance to live in one of those in which the conditions proved favorable for intelligent life. A combination of parameters that would be fantastically improbable in one universe would be probable somewhere among a large enough set of universes. The right combination would eventually come up by chance, like the winning combination on a Las Vegas slot machine.

Some cosmologists favor such theories partly because they do not have the theistic associations of a single Big Bang. Several such "many-worlds theories" have been proposed:[7]

1. *Successive Cycles of an Oscillating Universe.* Before the present era of expansion there could have been an era of contraction—a Big Crunch before the Big Bang. Any evidence for past cycles would have to be indirect, since their structure would have been totally wiped out in the fireball between cycles. The velocity of expansion is very close to the critical threshold between expanding forever (an *open* universe) and expanding a very long time before contracting again (a *closed* universe), but recent evidence favors an open universe with no future cycles.

2. *Multiple Isolated Domains.* Instead of multiple bangs in successive cycles, a single Big Bang might have produced multiple domains existing simultaneously. The domains would be like separately expanding bubbles isolated from each other because their velocity of separation would prevent communication even at the speed of light. They would be beyond our horizon of possible observation. The theory of isolated domains seems to be in principle untestable; support for it is at the moment based more on philosophical assumptions than on scientific evidence.

3. *Many-Worlds Quantum Theory.* In the next chapter we will examine Hugh Everett's proposal that every time there is a quantum indeterminacy in an atom, the universe splits into multiple branches, in each of which one of the potential outcomes is actualized. This interpretation of quantum theory involves a mind-boggling multiplicity of worlds, since each world would have to split again into many branches during each of the myriad atomic and subatomic events throughout time and space. This theory, too, is highly speculative and could not be directly verified, since no communication would be possible between the various branching worlds.

4. *Quantum Vacuum Fluctuations.* Quantum theory allows very brief violations of the law of conservation of energy, provided the debt is rapidly repaid. In the laboratory, a vacuum is really a sea of activity in which pairs of virtual particles

come into being and almost immediately annihilate each other again. One might think that the energy needed to make a universe would be enormous, but the gravitational energy is negative and would therefore reduce the total required. Some authors have presented the theory of quantum fluctuations as a secular version of creation *ex nihilo* (out of nothing), since the theory starts with a vacuum. But actually a vacuum is not "nothing," since the theory assumes the existence of a quantum field and the laws of quantum physics. How do we account for the situation within which one or more gigantic quantum fluctuations could occur? The atheist says we just have to assume it and treat it as a given.

Chance might play a role in any of these many-worlds scenarios. They all assume that our universe was once smaller than an atom is today. We will see (Chapter 3) that the quantum laws applicable at such subatomic dimensions do not permit precise predictions but only specify a range of probable values for all measurements. The precise value within any probability distribution seems to be a matter of chance. In different universes the values of the fundamental constants might differ. We just happen to be in one where life and intelligence are possible.

Of course, some of the parameters that seem to be arbitrary may turn out to be *necessary consequences* of more fundamental laws. In the history of science, many apparent coincidences later received theoretical explanation. Perhaps the constants can have only the values they have, and law rather than chance determines them. A few years ago, cosmologists could not explain the slight imbalance between particles and antiparticles in the early universe that led to the formation of matter as we know it. Now they think that the imbalance arose from a slight asymmetry in the decay processes of X particles and anti–X particles. Other "amazing coincidences" may be explainable in the future.

According to scientific materialists, cosmology leads us to *chance* or *necessity* but not to design or purpose. Peter Atkins defends the quantum fluctuation theory and argues that it is incompatible with theism.[8] Alan Guth subscribes to the quantum fluctuation theory and holds that the presence of "fine-tuned constants" in our universe was a matter of chance among many universes.[9] Some cosmologists have ended with a cosmic pessimism. Thus Steven Weinberg wrote in 1977 that humanity is alone in an "overwhelmingly hostile universe" headed for oblivion. Scientific activity, he said, is the only source of consolation in a meaningless world:

> The more the universe seems comprehensible, the more it also seems pointless. But if there is no solace in the fruits of research, there is at least some consolation in the research itself. . . . The effort to understand the universe is one of the very few things that lifts human life above the level of farce, and gives it some of the grace of tragedy.[10]

In a recent book Weinberg qualifies his earlier statement: "I did not mean that science teaches us that the universe is pointless, but only that the universe itself suggests no point."[11] The theist could then reply that if purpose is not ruled out by science, cosmic history can perhaps be coherently interpreted as an expression of God's purposes. A cosmic process producing intelligent persons is what we would expect if God is intelligent and personal. The conflict, then, is not really between science and religion but between alternative basic beliefs, materialistic and theistic. We will examine later how chance and necessity can be understood within a theistic framework.

2. Harmonizing Genesis and the Big Bang

When Galileo claimed that the sun rather than the earth is the center of the planetary system, he was condemned mainly because of his challenge to the authority of the church, though his challenge to a literal interpretation of scripture played a role. But by the seventeenth century Copernican astronomy was widely accepted. The scientific evidence was by then indisputable, and the metaphorical interpretation of scriptural passages that seemed to conflict with it was acknowledged in most Catholic and Protestant circles. In the nineteenth century the claim of biblical literalists that the universe is only a few thousand years old conflicted with the evidence from geology, evolutionary theory, and paleontology (the study of fossils), but not with astronomy. By the twentieth century, astronomy also provided evidence of a long cosmic history.

Gerald Schroeder, a physicist committed to Orthodox Judaism, argues that *the six days of Genesis* in God's time are *fifteen billion years* on our time-scale, since in relativity theory time measurements vary according to the frame of reference of the observer. The time dilation predicted by Einstein has been confirmed experimentally. For example, the average life of a meson (an unstable subatomic particle) traveling at nearly the speed of light around the racetrack of a laboratory accelerator is extended enormously, so it can make many more trips around the track than one would expect without the time dilation. Schroeder says that in a rapidly expanding universe, a day in God's time (identified with the whole cosmos) would be a few billion years for processes on earth. In the creation of Adam on the sixth cosmic day God was for the first time intimately linked to our planet. After that event, our time-scale was identified with God's, so the chronology of subsequent events in the Bible (including the lifespans of all Adam's descendants) is accurately recorded on a universal time-scale.[12]

Schroeder holds that other scientific facts can be found in later rabbinic writings. He describes in detail the commentary on Genesis by the thirteenth-century kabbalist Nahmanides, who said that creation started from an intangible substance of very small size, "no larger than a grain of mustard," from which the universe expanded and matter appeared. Nahmanides also said that there were ten principles or dimensions of reality corresponding to the ten times that the phrase "and God said" is repeated in Genesis. Schroeder claims that this has been confirmed in a remarkable way by recent superstring theory, which (as we saw) postulates ten initial dimensions.

Hugh Ross is an evangelical Christian with a Ph.D. in astronomy. One of his books is devoted to the ten dimensions of superstring theory and what he calls *the extra-dimensionality of God*. He cites many biblical passages in which God did things before the creation of the world—for example, Christ was chosen "before the foundation of the world" (1 Pet. 1:20). So Ross argues that God must operate from an additional time dimension, requiring eleven rather than ten dimensions in all. Within our normal experience, things are possible in three dimensions that could not occur in two. On a two-dimensional piece of paper, your pen tip at point A can reach point B only by traversing a series of points between; but with a third dimension, your pen can leave the paper at point A and suddenly reappear at B. So too, says Ross, God can do things with all those extra dimensions that would otherwise be impossible. Jesus could walk on water in defiance of gravity, for example, and he could walk through a closed door to talk to his disciples after the resurrection.[13]

According to Ross, events in our lives occur in the normal ("horizontal") temporal order in which causes precede effects. But *God's* time dimension is perpendicular to ours ("vertical") and intersects our whole series of temporal events simultaneously. For Ross, this provides a resolution of the long-standing paradox of free will and predestination. We make our choices freely in our own time-frame, but they are all included together in God's unified knowledge and action. God dwells in multiple time dimensions and empowers us in unexpected ways. "These discoveries about extra-dimensionality bear the potential to boost significantly Christians' awareness of the extent of power God has made available to us."[14]

I suggest that such arguments assume a literalistic interpretation of scripture that raises more problems than it solves. By seeking contemporary scientific theories hidden within a scriptural passage, they divert attention from its central religious message. Moreover, the use of superstring theory seems to me particularly dubious because it is highly abstract and speculative and cannot be tested experimentally at energies available in any existing or projected particle accelerator. We can indeed think of dimensions as differing ways of ordering a set of events from differing perspectives or in differing frameworks, and this may provide suggestive analogies for theology. But within theoretical physics the extra dimensions do not point to any supernatural forces beyond the world of space and time.

INDEPENDENCE

Science and religion cannot conflict if they are independent and autonomous enterprises, each asking a distinctive type of question, employing distinctive methods, and serving distinctive functions in human life. We will consider claims that the religious meaning of creation and the function of creation stories in human life are unrelated to scientific theories about physical events in the distant past.

I. The Religious Meaning of Creation

In "Separate Domains" (in Chapter 1) I summarized the neo-orthodox belief that the Bible should be taken *seriously but not literally*. Many theologians and biblical scholars share this view and claim that Genesis witnesses to a fundamental and enduring relationship between God and the world. It does so, they say, by telling a symbolic and poetic story that assumes the prescientific cosmology of its day. That cosmology included a short earth history, an earth-centered astronomy, and a three-decker universe (with heaven above and hell below our world)—ideas that are not consistent with modern science. But scripture conveys religious ideas that we can still accept, independent of any cosmology, ancient or modern. Genesis makes three theological affirmations: (1) the world is essentially good, orderly, and coherent; (2) the world is dependent on God; and (3) God is sovereign, free, transcendent, and characterized by purpose and will. These are characteristics of the world and God at every moment in time, not statements about events in the past.[15]

Within the Bible, the idea of *creation* does not refer simply to beginnings, because it is always linked to the idea of *redemption*. The formative events for Israel as a people were its liberation from oppression in Egypt and the covenant at Sinai. Most scholars hold that the first chapter of Genesis is a relatively late writing, probably from the fifth century B.C.E., in which the covenant faith was given a more universal context, and God was worshiped as Lord of nature as well as history.[16] Isaiah ties past, present, and future together; God is indeed the creator of Israel, but also of all humanity and all nature. In the future God will again liberate a people in bondage and bring a new harmony to all nations and all nature (Isa. 40, 45, 49). In the New Testament, too, creation is inseparable from redemption. The opening verse of John's Gospel recalls Genesis: "In the beginning was the Word, and the Word was with God. . . . All things were made through him." Here the term *Word (logos)* merges the Greek principle of rationality with the Hebrew idea of God's Word active in the world. But John then ties creation to revelation: "And the Word became flesh." In Christ's life and death, according to the early church, God made known the purpose of creation.

The idea of creation *ex nihilo* (out of nothing) is not stated in Genesis. The opening verses start with a watery chaos: "The earth was without form and void, and darkness was upon the face of the deep and the Spirit of God was moving over the face of the waters." But by the fourth century the Christian community faced rival philosophies, and it formulated the *ex nihilo* doctrine in response to them. Against the Gnostic teaching that matter is evil—the work of a lesser being and not the work of the God who redeems—the *ex nihilo* doctrine said that the Creator and Redeemer were one. Against claims that preexisting matter limited God's creativity, it asserted that God is the source of matter as well as form. Against pantheism, it asserted that the world is not divine or part of God but is distinct from God. Against the idea that the world was an emanation of God, made of the divine substance and sharing its characteristics, the *ex nihilo* doctrine asserted that God is transcendent and essentially different from the world. It was such theological assertions, and not any specific reference to a temporal beginning, that were important in the early church's self-definition.

In the fourth century Augustine was willing to accept *metaphorical* or *figurative* interpretations of Genesis, and he said that it was not the intent of scripture to instruct us about such things as the form and shape of the heavens. "God did not wish to teach men things not relevant to their salvation." He held that creation is not an event in time; time was created along with the world. Creation is the timeless act through which time comes to be and the continuous act by which God preserves the world. He said that it is meaningless to ask what God was doing before creating the world, for there was no time without the created world.[17]

Thomas Aquinas in the thirteenth century accepted a beginning in time as part of scripture and tradition and said that creation in time helps to make God's power evident. But he argued that a universe that had always existed would equally require God as Creator and Sustainer. What is essential theologically could be stated *without reference to a beginning* or a singular event. To be sure, one of the versions of his cosmological argument did assume a beginning in time: every effect has a cause, which in turn is the effect of a previous cause, back to a First Cause that initiated the causal chain. But in another version, he asks, Why is there anything at all? He replies that the whole causal chain, whether finite or infinite, is dependent on God. God's priority is metaphysical rather than temporal.

There has also been a subordinate theme of *continuing creation* from biblical times to the present. Edmund Jacob has said that while there are many biblical texts referring to a primordial creation in the beginning, "other texts, generally more ancient, draw much less distinction between the creation and conservation of the world, and make it possible for us to speak of a *creatio continua*."[18] God is still creating through natural processes. "Thou dost cause the grass to grow for cattle and the plants for man to cultivate. . . . When thou sendest forth thy Spirit, they are created; and thou renewest the face of the ground" (Ps. 104:14, 30).

Should we take a *beginning of time* literally, even if we do not interpret the six days in Genesis literally? Here theologians are divided. For one thing, the biblical concept of finite linear time has contributed to the Western view of history. The West has differed from the ancient cultures and Eastern religions, which have assumed an infinite succession of cycles; these cultures have generally evidenced less interest in historical development. But other theologians suggest that even a beginning of time is not crucial to the theological notion of creation. David Kelsey, for instance, says that the basic experience of gratitude for life as a gift has no essential connection with speculations about unique events at the beginning. Science and religion, he maintains, address different questions, and those questions should not be confused.[19] He defends a strong version of the Independence thesis.

2. The Function of Creation Stories

In Chapter 1 (see "Differing Languages") we noted that linguistic analysts claim creation stories serve functions in human life that are very different from those of scientific theories. Anthropologists and historians of the world religions have pointed out that people in all cultures have sought to locate their lives within a cosmic order. Human interest in origins may be partly speculative or explanatory, but it is driven mainly by the need to understand who we are in a larger framework of meaning and significance. Creation stories provide patterns for human behavior and archetypes of authentic human life in accord with a universal order. They portray basic relationships between human life and the world of nature. Often they acknowledge structures of integration and creativity over against powers of disintegration and chaos.

The *Babylonian creation story*, which is earlier than Genesis, also starts with a primeval watery chaos. One of its characters is the sea monster Rahab, who is mentioned in several biblical passages. Many texts in the Hebrew scriptures assume a continuing struggle between order and chaos, and they acknowledge the presence of evil and the fragility of creation.[20] But the biblical story differs from other ancient creation stories in its assertion of the sovereignty and transcendence of God and the dignity of humanity. God is portrayed as purposive and powerful, creating by word alone. Creation is orderly and deliberate, following a comprehensive plan and resulting in a harmonious and interdependent whole. In the Babylonian story, humanity was created to provide slaves for the gods; in Genesis, humanity was given a special status in God's plan, superior to the rest of creation.[21] The Genesis account differs from the Babylonian account in the beliefs it expresses, but it serves the same functions and meets the same human needs.

A religious community appropriates and participates in its *sacred stories* in various ways. Often the stories are symbolized or enacted in rituals. Frederick Streng speaks of each generation passing on the stories that "manifest the essential structure of reality." Mircea Eliade says that exemplary patterns in primordial time are made present in ritual and liturgy.[22] Several early biblical psalms suggest a ceremony celebrating Israel's life in relation to God as Creator (Psalms 93 and 99). The traditional Jewish morning prayer uses the present tense:

> *Praised are You, O Lord our God, King of the universe.*
> *You fix the cycles of light and darkness;*
> *You ordain the order of all creation. . . .*
> *In Your goodness the work of creation*
> *Is continually renewed day by day.*[23]

The idea of creation can also be seen as an expression of enduring *human experiences*, such as: (1) a sense of dependence, finitude, and contingency; (2) a response of wonder, trust, gratitude for life, and affirmation of the world; and (3) a recognition of interdependence, order, and beauty in the world. The religious idea of creation starts from wonder and gratitude for life as a gift. Theological doctrines are an attempt to interpret such experiences within the context of a particular historical tradition. The theological meaning of creation can be combined with a variety of physical cosmologies, ancient or modern, and does not require any one cosmology.

Proponents of the Independence thesis thus hold that the questions asked by the astronomer and those posed by the theologian are very different. Each mode of inquiry is selective and has its limitations. The language of science and the language of religion serve *differing functions* in human life, as linguistic analysts and anthropologists have pointed out. The goal of science is to understand lawful relations among natural phenomena, while that of religion is to follow a way of life within a larger framework of meaning.

A good case can be made for the Independence model. Because it sees the two fields as separate and independent enterprises, it rules out any possibility of conflict between them. But as I have noted, the price paid for this separation into watertight compartments is that positive relationships and coherent synthesis are also ruled out.

DIALOGUE

Advocates of Dialogue hold that science has presuppositions and raises limit-questions that science itself cannot answer. Religious traditions can suggest possible answers to such questions, these thinkers assert, without violating the integrity of science. The distinction between the disciplines is maintained, but thoughtful dialogue can occur. Two topics of dialogue arising from cosmology are the intelligibility and the contingency of the universe.

1. The Intelligibility of the Cosmos

Physicists and astronomers are seeking a unified theory, which has been called the Holy Grail of their current quest. The search for a unified theory is partly motivated by the conviction that the cosmos is orderly, simple, and *rationally intelligible*. Scientists must, of course, check their theories against experimental evidence, but they are convinced that a valid general theory will be conceptually simple and aesthetically beautiful. Einstein said that the only thing that is incomprehensible about the world is that it is comprehensible.

Historically, the conviction that the cosmos is unified and intelligible had both Greek and biblical roots. The Greeks, and later the Stoics in the Roman world, saw the universe as a single system. The Greek philosophers had great confidence in the power of reason, and it is not surprising that they made significant progress in mathematics and geometry. We saw in Chapter 1 that historians have claimed that the biblical doctrine of creation made a distinctive contribution to the rise of experimental science because it combined the ideas of *rationality* and *contingency*. If God is rational, the world is orderly; but if God is also free, the world did not have to have the particular order that it has. The world can then be understood only by observing it, rather than by trying to deduce its order from necessary first principles, as the Greeks tried to do. The church fathers said that God voluntarily created form as well as matter *ex nihilo*, rather than imposing preexisting eternal forms on matter.[24]

Thomas Torrance has written extensively on the theme of *"contingent order."* He stresses God's freedom in creating as an act of voluntary choice. God alone is infinitely free, he says, and both the existence and the structure of the world are contingent in the sense that they might not have been. Furthermore, the world might have been differently ordered. We can discover its order only by observation, not by logic. Moreover, the world can be studied on its own, because in being created it received its own independent reality, distinct from the transcendent God. Science can legitimately examine the world apart from God, while the theologian can still assert that the world is ultimately dependent on God.[25]

Albert Einstein, on the other hand, saw any contingency as a threat to belief in the rationality of the world, which he said is central in science. "A conviction, akin to religious feeling, of the rationality or intelligibility of the world lies behind all scientific work of a high order."[26] He spoke of a "cosmic religious sense" and "a deep faith in the rationality of the world." He rejected the idea of a personal God whose acts arbitrarily interfere in the course of events; he subscribed to a form of pantheism, identifying God with the orderly structure itself. When asked if he believed in God, he replied, "I believe in Spinoza's God, who reveals himself in the orderly harmony of what exists."[27] Einstein equated rationality with orderliness and determinism; he never abandoned his conviction that the uncertainties of quantum theory only reflect temporary human ignorance, which will be left behind when the deterministic underlying mechanisms are discovered. He was mainly concerned about the necessity of events, but he also thought that the laws of physics are logically necessary.

In *The Moment of Creation*, the physicist James Trefil describes the search for unified laws in cosmology. In an epilogue he writes:

> But who created those laws? . . . Who made the laws of logic? . . . No matter how far the boundaries are pushed back, there will always be room both for religious faith and a religious interpretation of the physical world. For myself, I feel much more comfortable with the concept of a God who is clever enough to devise the laws of physics that make the existence of our marvelous universe inevitable than I do with the old-fashioned God who had to make it all, laboriously, piece by piece.[28]

John Polkinghorne discusses the intelligibility of the world in a theistic framework. The key to understanding the physical world, he claims, is mathematics, an invention of the human mind. The fit between rationality in our minds and rationality in the world is to be expected if the world is the creation of mind. God is *the common ground of rationality* in our minds and in the world. Orderliness can also be understood as God's faithfulness, but it does not exclude an important role for chance. Polkinghorne invokes the early Christian concept of *logos*, which, as we have seen, combined the Greek idea of a rational ordering principle and the Hebrew idea of the active Word of God. He maintains that the theist can account for the intelligibility that the scientist assumes.[29]

2. The Contingency of the Cosmos

Four kinds of contingency characterize the universe described by modern cosmology:[30]

1. *Contingent Existence.* Why is there anything at all? This is the question of greatest interest to theologians, and it corresponds to the religious meaning of creation *ex nihilo* as indicated above. The existence of the cosmos as a whole is not self-explanatory, regardless of whether it is finite or infinite in time. It now appears likely that the Big Bang was an absolute beginning, a singular event, but if scientists were to discover new evidence for a cyclic universe or infinite time, the contingency of existence would remain. The details of particular scientific

cosmologies are irrelevant to the contingency of the world. Even if a theory were to show that there is only one possible universe, the universe would still remain only possible; nothing in the theory would provide that a universe actually exists or that the theory is instantiated. Stephen Hawking writes:

> Even if there is only one possible unified theory, it is just a set of rules and equations. What is it that breathes fire into the equations and makes a universe for them to describe? The usual approach of science of constructing a mathematical model cannot answer the questions of why there should be a universe for the model to describe.[31]

2. *Contingent Boundary Conditions.* If it turns out that past time was finite, there was indeed a singularity, inaccessible to science. Such a beginning was assumed by the church fathers in the classical *ex nihilo* doctrine, even though it was not their chief concern. As Aquinas said, such a beginning would provide an impressive example of dependence on God. On the other hand, if it turns out that time is infinite, we would still have contingent boundary conditions; however far back we went, we would have to deal with a situation or state that we would have to treat as an unexplained given.

Hawking has developed a theory of quantum gravity that assumes neither infinite time nor a beginning of time. Instead, time is paradoxically *finite but unbounded*. His equations for the early universe involve imaginary time, which is indistinguishable from the three spatial dimensions. Just as the two-dimensional surface of the earth is finite but has no edges, and three-dimensional relativistic ("curved") space is finite but unbounded, so Hawking's spatial and imaginary time dimensions are all finite but unbounded. In that imaginary time-frame, real time gradually emerges.[32] Hawking grants that the interpretation of events in imaginary time is not clear, and some critics say that it is inconsistent to speak of time as *emerging*, because emergence refers to changes in real time. I suggest that even if time in its early stages was "fuzzy" and had no sharp edge that could be called a "beginning," the theory postulates a given structure of physical relationships from which space and time develop.

3. *Contingent Laws.* While many of the laws of cosmology appear to be arbitrary, some of them may turn out to be necessary implications of more fundamental theories. If a unified theory is found, however, it will itself be contingent, and the argument only moves back a stage. Moreover, some laws applicable to higher emergent levels of life are not derivable from the laws of physics. In later chapters I will suggest that life and consciousness cannot be explained by the laws of physics, though they do not violate physical laws. It is misleading to refer to a unified theory in physics as a Theory of Everything, for its unity would be achieved only by a very high degree of abstraction that would leave out all of the diversity and particularity of events in the world and the emergence of more complex levels of organization from simpler ones. We could hardly expect a unified theory to tell us very much about an amoeba, for example, much less about Shakespeare, Beethoven, or Newton.

4. *Contingent Events.* In the next chapter I will argue that uncertainty in quantum physics reflects indeterminacy in the world and not simply the limitations of our knowledge. The cosmos is a unique and irreversible sequence of events in which genuine novelty appears. Our account of it must take a historical form rather than consisting of general laws alone. We will encounter a similar contingency in the mutations, gene recombinations, and environmental conditions that influenced evolutionary history.

Of course, many scientists today try to confine themselves to strictly scientific questions, or they may adopt an explicitly materialistic philosophy. Yet wider reflection on cosmology seems to be an important way of raising what I have called limit-questions. At the personal level, astronomers often express a sense of mystery and awe at the power unleashed in the Big Bang and the occurrence of phenomena at the limits of our experience, language, and thought. Cosmology encourages the examination of our presuppositions about time and space, law and chance, necessity and contingency. Above all, the intelligibility of the cosmos suggests questions that arise in science but cannot be answered within science itself.

INTEGRATION

Proponents of the Integration thesis seek a closer correlation of particular religious beliefs with particular scientific theories than is advocated by exponents of Dialogue. The Anthropic Principle can be interpreted as a new version of natural theology that starts from cosmology. A theology of nature is represented here by models of God as Creator that express the central beliefs of the religious community but incorporate theological reformulations in response to current cosmology. We close by looking at the theological significance of humanity in the light of recent cosmology.

1. Design: The Anthropic Principle

A striking feature of the new cosmological theories is that even a small change in the physical constants would have resulted in an uninhabitable universe. Among the many possible universes consistent with Einstein's equations, ours is one of the few in which the arbitrary parameters are just right for the existence of anything resembling organic life. Many cosmologists have pointed out that the possibility of life as we know it depends on the value of a few basic constants and is remarkably sensitive to them.[33] Among these fine-tuned phenomena are the following:

1. *The Expansion Rate.* Stephen Hawking writes, "If the rate of expansion one second after the Big Bang had been smaller by even one part in a hundred thousand million it would have recollapsed before it reached its present size."[34] On the other hand, if it had been greater by one part in a million, the universe would have expanded too rapidly for stars and planets to form. The expansion rate itself depends on many factors, such as the initial explosive energy, the mass of the universe, and the strength of gravitational forces. The cosmos seems to be balanced on a knife edge.

2. *The Formation of the Elements.* If the strong nuclear force had been even slightly weaker, we would have had only hydrogen in the universe. If the force had been even slightly stronger, all the hydrogen would have been converted to helium. In either case, stable stars and compounds such as water could

not have been formed. Similarly, the nuclear force is only barely sufficient for carbon to form; yet if it had been slightly stronger, the carbon would all have been converted into oxygen. The element carbon has many properties that are crucial to the later development of organic life as we know it.

3. *The Particle/Antiparticle Ratio.* For every billion antiprotons in the early universe, there were one billion and one protons. The billion pairs annihilated each other to produce radiation, with just one proton left over. A greater or smaller number of survivors—or no survivors at all (if they had been evenly matched)—would have made our kind of material world impossible. If the laws of physics are symmetrical between particles and antiparticles, why was there a tiny asymmetry here?

The simultaneous occurrence of many independent improbable features appears *wildly* improbable. Reflection on the way the universe seems to be fine-tuned for intelligent life has led some cosmologists to formulate the *Anthropic Principle:* what we can expect to observe must be restricted by the conditions necessary for our presence as observers.[35] The principle underscores the importance of the observer in science, but it does not in itself provide any causal explanation of the presence of observers. However, this fine-tuning could be taken as an argument for the existence of a designer, perhaps a God with an interest in conscious life.

Some physicists see *evidence of design* in the early universe. Stephen Hawking, for example, writes, "The odds against a universe like ours emerging out of something like the Big Bang are enormous. I think there are clearly religious implications."[36] And Freeman Dyson, in a chapter entitled "The Argument from Design," gives a number of examples of "numerical accidents that seem to conspire to make the universe habitable." He concludes, "The more I examine the universe and the details of its architecture, the more evidence I find that the universe in some sense must have known we were coming."[37]

We have seen that critics of the Anthropic Principle hold that the presence of the right combination of constants may have been *a matter of chance* among a large number of universes (in most of which life would have been impossible). Other critics speculate that all of these apparently arbitrary constants can be derived from one all-inclusive equation not yet known to science. Perhaps that equation will show that the only possible universe is one with constants just like ours. This would mean that if there is a universe at all its constants are necessary and not the product of chance. The theist can reply that such a theory would only push the argument back a stage. For it is all the more remarkable if a highly abstract physical theory, which itself has absolutely nothing to say about life, turns out to describe structures that have the potential for developing into life. The theist could welcome this as part of God's design. A theory that starts with a single equation would still leave unanswered the question, Why *that* equation?[38]

In short, if chance and necessity are possible alternatives to design, the Anthropic Principle does not provide a conclusive argument for the existence of God of the sort once sought in natural theology. However, the principle is quite consistent with *a theology of nature* in which belief in God rests primarily on other grounds. The fine-tuning of the physical constants is just what one would expect if life and consciousness were among the goals of a rational and purposeful God.

2. Models of God as Creator

In the Christian community, belief in God rests primarily on the historical witness to redemption in the covenant with Israel and the person of Christ and on the personal experience of forgiveness and renewal. The doctrine of creation represents the extension of these ideas of redemption to the world of nature. We have said that it also expresses the experience of wonder, dependence on God, gratitude for life as a gift, and recognition of order and novelty in the world. As advocates of Independence hold, these beliefs do not require particular cosmological theories, ancient or modern. Nevertheless, there are points of overlap and there are areas of potential conflict in which traditional beliefs may need to be reformulated.

I have argued that in talking about unobservable entities in both science and religion, we have to use *imaginative models*. Sometimes we need to use more than one model—wave and particle models of an electron, for example, or personal and impersonal models of God. Models are conceptually elaborated in scientific theories and in theological doctrines, and these theories and doctrines have to be tested against the experience of a community (scientific or religious). The testing process is more rigorous in the case of science, but theological beliefs are indeed reformable in the course of time. Models in both fields are formed on the basis of metaphors or analogies drawn from more familiar situations.[39]

In the Bible, *models of God as Creator* are offered at many points. In Genesis, God is a purposeful designer imposing order on chaos. God's command is powerful, and the divine Word is effective. Other biblical images picture a potter forming an object (Jer. 18:6, Isa. 64:8) or an architect laying out the foundations of a building (Job 38:4). God is Lord and King, ruling the universe to bring about intended purposes. The world is a manifestation of God's Word and an expression of divine Wisdom, which communicates meaning. In the New Testament, God creates through the Word (John 1), a term that, as we have seen, brings together the Hebrew concept of divine Word active in the world and the Greek concept of Word *(logos)* as rational principle. The purpose of creation is made known in Christ, the Word incarnate. Here is a rich diversity of models, each a partial and limited analogy, highlighting imaginatively a particular way of looking at God's relation to the world.

The potter and craftsman analogies assume the production of a completed, static product. They seem less helpful in thinking about an ongoing, dynamic, evolutionary process. The image of God as gardener is more promising, though it occurs rarely in the Bible (for example, Gen. 2:8), perhaps because the Israelites wanted to distance themselves from the nature gods of surrounding cultures. The analogies of God as King and Ruler were emphasized in medieval and Calvinist thought. But the doctrines of omnipotence and predestination to which they led are difficult to reconcile with the current scientific view of nature, as we shall see later.

In the Bible, the model of father is used for God's relation to persons. God as mother was a rare image in a patriarchal society, but it appears occasionally (for example, Isa. 49:15 and 66:13). The parental analogy is usually drawn from a parent nurturing a growing child rather than from procreation and birth. This seems a particularly appropriate image of God's relation to the world; the wise parent allows an increasing independence in the child while offering encouragement and love. Such an image can maintain a balance between what our culture thinks of as masculine and feminine qualities, in contrast to the heavily "masculine" monarchial model of omnipotence and sovereignty.

The biblical model of God as Spirit seems to me particularly help-ful. In the opening verses of Genesis, the Spirit of God moves over the face of the water. The Spirit is active in the continuing creation of plants and animals, expressed in the present tense: "When thou sendest forth thy Spirit, they are created" (Ps. 104:30). The Spirit also represents God's activity in the worshiping community and in the inspiration of the prophets, and the Spirit descends on Christ at his baptism and on his followers at Pentecost. Reference to the Spirit thus ties together God's work as Creator and as Redeemer.

The biblical concept of divine Word *(logos)* suggests an analogy taken from contemporary science. The *communication of information* is an important concept in communication theory, computer net-works, and the DNA in organisms. In each case communication requires selective response (decoding) and the interpretation of a message in a wider context. Several authors, including John Polkinghorne, have proposed a model of God as communicator of information that does not require the violation of scientific laws. The biblical idea of the divine Word can be viewed as the communication of rational structure and meaning when the world is interpreted in a wider context. These authors use science not as a proof of God's exis-tence but as a source of new analogies for talking about God. We will explore these and other models further in Chapters 4 and 6.

3. The Significance of Humanity

Much of the resistance to Copernicus and Galileo arose because in their cosmologies the earth, no longer the center of the universe, was only one of several planets going around the sun. Darwin carried fur-ther the demotion of humanity from its central place in the cosmic scheme and challenged the biblical understanding of the significance of human life. What are the implications of current cosmology for our self-understanding? Can they be reconciled with the biblical view of humanity?

1. *The Immensity of Space and Time.* Humanity seems insignifi-cant in the midst of vast stretches of time and space. But today those immensities do not seem inappropriate. We now know that it takes about fifteen billion years for heavy elements to be "cooked" in the interior of stars and then

scattered to form a second generation of stars with planets, followed by the evolution of life and consciousness. A very old expanding universe has to be a huge universe—on the order of fifteen billion light-years. However, as Teilhard de Chardin pointed out, we should not measure significance by size and duration, but by such criteria as complexity and consciousness.[40] The greatest complexity is present in the middle range of size, not at atomic dimensions or galactic dimensions. There are trillions of neurons in a human brain; the number of possible ways of connecting them is greater than the number of atoms in the universe. A higher level of organization and a greater richness of experience occur in a human being than in a thousand lifeless galaxies. It is human beings, after all, who reach out to understand that cosmic immensity.

2. *Interdependence.* Cosmology joins evolutionary biology, molecular biology, and ecology in showing the interdependence of all things. We are part of an ongoing community of being; we are kin to all creatures, past and present. From astrophysics we know about our indebtedness to a common legacy of physical events. The chemical elements in your hand and in your brain were forged in the furnaces of stars. The cosmos is all of a piece. It is multileveled; each new, higher level was built on lower levels from the past. Humanity is the most advanced form of life we know, but it is part of a wider process in space and time. The new view may undercut claims that set humanity completely apart from the rest of nature, but it by no means makes human life insignificant.

3. *Life on Other Planets.* Planets are so numerous that if even a small fraction of them are habitable, life could exist in many stellar systems. Observations with the Hubble telescope in 1999 found evidence of another star with more than one planet. Extraterrestrial intelligence has been the subject of many science fiction films, from *E.T.* and *Star Trek* to *Contact.* The possibility of beings superior to us, living in more advanced civilizations, is a further warning against human pride. It also calls into question exclusive claims concerning God's revelation in Christ. Here we can recall that

even on our planet the work of the *logos*, the eternal Word, was not confined to its expression in Christ. If that Word is active in continuing creation throughout the cosmos, we can assume that it will also have revealed itself as the power of redemption at other points in space and time, in ways appropriate to the forms of life existing there. Our images of God must be commensurate with the cosmic scale of creation and redemption.

4. *Chance and Purpose.* Traditionally, God's purpose in creation was identified with order. An emphasis on God's sovereignty led to a divine determinism in which everything was thought to happen in accordance with a detailed plan; any element of chance was viewed as a threat to God's total control. It is not surprising that some scientists and philosophers impressed by the role of chance have been led to reject theism. They view life as the accidental result of chance, and they assume that chance and theism are incompatible. Whereas the appropriate response to design would be gratitude and thanksgiving, the response to pure chance would be a sense of futility and cosmic alienation.

One possible answer is to say that God really controls all the events that appear to us to be chance—whether in quantum uncertainties, evolutionary mutations, or the accidents of human history. This would preserve divine determinism at a subtle level undetectable to science. But I will argue in Chapter 4 that the presence of genuine chance is not incompatible with theism. We can see design in the whole process by which life came into being, with whatever combination of probabilistic and deterministic features the process had. Natural laws and chance may equally be instruments of God's intentions. There can be purpose without an exact predetermined plan.

Proponents of a theology of nature thus draw extensively from a historic tradition and a worshiping community, but they are willing to modify some traditional assertions in response to the findings of science. In the case of cosmology this must be done with caution, since many current theories are highly speculative and evidence is at best very indirect. At the moment, a singular Big Bang seems the most plausible theory, and the theist can see it as a moment of divine initiation, but we should not tie our religious beliefs irrevocably to any one theory.

Looking at pictures of the earth taken by astronauts on the moon, we can celebrate the beauty of our amazing planet and express gratitude for the gift of life. Standing under the stars at night, we can still experience wonder and awe. Now we know that the cosmos has included stretches of space and time that we can hardly imagine. What sort of world is it in which those strange early states of matter and energy could be the forerunners of intelligent life? Within a theistic framework it is not surprising that there is intelligent life on earth; we can see here the work of a purposeful Creator. Theistic belief makes sense of this datum and a variety of other kinds of human experience, even if it offers no conclusive proof. We still ask: Why is there anything at all? Why are things the way they are? With the psalmist of old we can say, "O Lord, how manifest are thy works! In wisdom thou has made them all. . . . When thou sendest forth thy Spirit, they are created" (Ps. 104:30).

Chapter Three

The Implications of Quantum Physics

*P*hysics is the study of the basic structures and processes of change in matter and energy. Dealing with the lowest organizational levels, and using the most abstract equations, it seems of all sciences the furthest removed from the concern of religion for human life. But physics is of great historical and contemporary importance because it was the first science that was systematic and exact, because many of its assumptions were taken over by other sciences, and because it has also exerted a major influence on philosophy and theology.

Moreover, though physicists study only the inanimate, they now look at very diverse objects, from quarks and atoms to solid-state crystals, planets, and galaxies—including the physical basis of life processes. Without the forces studied by quantum physics, we would not have chemical elements, the periodic table, transistors, nuclear power, or life itself. The ideas of quantum physics are not easy to understand, but they repay careful study because they raise basic issues of the relations between law and chance, part and whole, the observer and the observed. The idea of indeterminacy in quantum theory has been particularly prominent in recent discussions of science and religion. The other major revolution in twentieth-century physics, Einstein's theory of relativity, has important implications for our understanding of space, time, energy, and matter, but I will mention it only briefly in order to avoid making this chapter too long. My own conclusions about quantum theory are summarized at the end of the chapter.

Newton held that the world consists of particles in motion, and this view was widely adopted in the eighteenth century. With the rise of chemistry in the nineteenth century, these particles were identified with the atoms of Dalton's theory, but the assumptions of Newtonian physics were still accepted by most scientists. The Newtonian outlook was *deterministic* in holding that, in principle, the future course of any system could be predicted from accurate knowledge of its present state. It was *reductionistic* in holding that the behavior of a system is determined by the behavior of its smallest parts. And it was *realistic* in assuming that scientific theories describe the world as it is in itself, apart from the involvement of the observer. All three of these assumptions have been challenged by quantum physics in the twentieth century.

Early in the century, several experiments suggested that light travels in separate wave packets called *quanta*. Niels Bohr's first model of the atom pictured electrons following orbits around the nucleus like planets in a miniature solar system (as pictured still in popular images of the atom). The model could not account for the discrete energy levels of the orbital electrons, but it could account for the frequencies of the light given off when an electron returned to one of the inner orbits after it had been excited to an outer orbit. The emitted light behaved very much like a stream of particles with definite energy and momentum. In other experiments, a stream of electrons, which had always been viewed as localized particles, showed the spread-out interference patterns characteristic of waves.

During the 1920s, Erwin Schrödinger formulated the *wave equation* of quantum theory, which accounts accurately for the discrete energy levels of electrons in atoms. But the equation does not predict the location or trajectory of an electron. It permits only the calculation of the *probability* that an electron will show up at a particular point when an observation is made. The equation portrays a complex pattern of probability waves that are accurately correlated with a large set of observations but not with an individual observation. The electrons in the Bohr model had been readily visualizable, but the electrons in quantum theory could not be pictured at all. One might try to imagine them as patterns of probability waves filling the space around the nucleus like the vibrations of a three-dimensional symphony of musical tones of incredible complexity, but the analogy would not help us much. The quantum atom is inaccessible to direct observation and unimaginable in terms of everyday properties; it cannot even be coherently described in terms of classical physical concepts such as space, time, and causality.[1]

Quantum theory does not ascribe exact values to properties such as the position and momentum of the inhabitants of the atomic world between observations. It specifies only a range of values and the probability of individual values within that range. The *Heisenberg Uncertainty Principle* states that the more accurately we determine the position of an electron or other particle, the less accurately we can determine its momentum, and vice versa. A similar uncertainty relation connects other pairs of variables, such as energy and time. For example, we can predict when half of a large group of radioactive atoms will have disintegrated, but we cannot predict when a particular atom will disintegrate. We can predict only the probability that it will disintegrate in a given time interval; it might disintegrate in the next second, or in a thousand years. The extensions of quantum theory in recent years into the nuclear and subnuclear domains have maintained the probabilistic character of the earlier theory. *Quantum field theory* is a generalization of quantum theory that is consistent with Einstein's theory of relativity. It has been applied with great success to electroweak theory and subnuclear interactions (quantum chromodynamics or quark theory).

Three differing interpretations of these quantum uncertainties have been given by physicists, and each has important theological implications that we will explore in this chapter.

1. *Uncertainty as Human Ignorance.* A minority of physicists, including Einstein and Max Planck, have maintained that the uncertainties of quantum theory are attributable to our present ignorance. They believe that detailed subatomic mechanisms are rigidly causal and deterministic; someday the laws of these mechanisms will be found and exact prediction will be possible. Einstein wrote, "The great initial success of quantum theory cannot convert me to believe in that

fundamental game of dice. . . . I am absolutely convinced that one will eventually arrive at a theory in which the objects connected by laws are not probabilities but conceived facts."[2] Einstein expressed his own faith in the order and predictability of the universe, which he thought would be marred by any element of chance. "God does not play dice," he said.

More recently, David Bohm has tried to preserve determinism by constructing a new formalism with hidden variables at a lower level. The apparent randomness at the atomic level would arise from variations in the concurrence of exact forces among the hidden variables.[3] So far his calculations have yielded no predictions differing from those of quantum theory. Most scientists are dubious about such proposals. Unless someone can actually develop a better alternative theory that can be tested, they say, we had better accept the probabilistic theories we have and give up our nostalgia for the certainties of the past.

2. *Uncertainty as Experimental or Conceptual Limitations.* Many physicists assert that uncertainty is not a product of temporary ignorance but a fundamental limitation permanently preventing exact knowledge of the atomic world. The first version of this position, found in the early writings of Niels Bohr, claims that the difficulty is an experimental one; the uncertainty is introduced by *the process of observation.* If we want to observe an electron, we must bombard it with a quantum of light, which disturbs the situation we were attempting to study.[4] Although this interpretation fits many experiments, it appears unable to account for uncertainties when nothing is done to disturb the system—for example, the unpredictability of the time at which a radioactive atom spontaneously disintegrates.

The second version of the argument attributes uncertainty to our *inescapable conceptual limitations.* By our choice of experimental situations we decide in which of our conceptual schemes (wave or particle, exact position or exact momentum) an electron will manifest itself to us. We must choose either *causal* descriptions (using probability functions that themselves evolve deterministically) or *spatiotemporal* descriptions (using localized variables that are only statistically connected); we cannot have both at once. This interpretation is *agnostic* as to whether the atom itself, which we can never know, is determinate or indeterminate.

3. *Uncertainty as Indeterminacy in Nature.* In his later writings, Werner Heisenberg held that indeterminacy is *an objective feature of nature* and not a limitation of human knowledge.[5] Instead of assuming that an electron has a precise position and velocity that are unknown to us, we should conclude that it is not the sort of entity that always has such properties. Observing consists in extracting from the existing probability distribution one of the many *possibilities* it contains. The influence of the observer, in this view, does not consist in disturbing a previously precise though unknown value, but in forcing one of the many existing potentialities to be actualized.

If this interpretation is correct, indeterminacy characterizes the world. Heisenberg calls this "the restoration of the concept of potentiality." In the Middle Ages the idea of *potentiality* referred to the tendency of an entity to develop in a particular way. Heisenberg does not accept the Aristotelian idea that entities strive to attain a future purpose, but he does suggest that the probabilities of modern physics refer to tendencies in nature that include *a range of possibilities*. The future is not simply unknown; it is "not decided." More than one alternative is open, and there is some opportunity for unpredictable novelty. Time involves a unique historicity and unrepeatability; the world would not repeat its course if it were restored to a former state, for at each point a different event from among the potentialities might be actualized.

A more exotic version is Hugh Everett's *many-worlds interpretation*. As we noted in the previous chapter, Everett has proposed that every time a quantum system can yield more than one possible outcome, the world splits into many separate worlds, in each of which one of these possible outcomes occurs.[6] We happen to be in the world in which the outcome that we observe occurs, and we have no access to the other worlds in which duplicates of us observe other possibilities. The theory seems to be in principle untestable, since we have no access to other worlds containing the potentialities unrealized in ours. It seems much simpler to assume that potentialities not actualized in our world are not actualized anywhere. On that assumption, there is one universe that is objectively indeterminate.

We may conclude from this analysis that adherents of the second and third of these basic positions—which between them include by far the majority of contemporary physicists—agree in rejecting the determinism of Newtonian physics. What is the theological significance of this conclusion? Let us consider each of our four basic views.

CONFLICT

Biblical literalism did not lead to conflicts with Newtonian physics like those that occurred with Copernican astronomy or Darwinian evolution. But in the centuries following Newton, the idea of a universe rigidly determined by natural laws seemed incompatible with traditional ideas of God's action in the world. More recently, the role of chance in quantum phenomena has challenged ideas of divine purpose and sovereignty.

The conflicts with religion arising in the history of physics may not have been as dramatic as those arising from astronomy or evolution, but they have been important because physics has been viewed as the most fundamental of the sciences. The most significant conflict has involved the relation between God's control of events, determination by natural laws, and the presence of chance at the quantum level.

1. God in a Deterministic World

Newtonian physics did not at first seem to present any challenge to religious beliefs. Almost all of the scientists of the seventeenth century were devout Christians. Newton himself believed that God keeps the stars from collapsing under gravitational attraction and intervenes periodically to correct planetary perturbations in the solar system (perturbations that Pierre Laplace later showed would eventually cancel each other out). But Newton and his contemporaries saw God's hand mainly in the harmonious design of the universe. In their view, the world is an intricate machine following immutable laws, but it expresses the wisdom of an intelligent creator.[7]

Moreover, God was held to have an ongoing role in a *law-abiding world*. God not only designed the laws but sustains them continually. The laws are a continuing expression of God's purposes and sovereignty. God predestines and foresees all events; everything that happens is in accordance with God's will. Newtonian scientists held that the passivity of matter in their science is consistent with traditional ideas of God's sovereignty and transcendence and is inconsistent with pantheism, astrology, vitalism, and alchemy, in which matter itself was held to be more active. Newtonians agreed with medieval philosophers that God as "primary cause" makes use of the "secondary causes" described by science as instruments to achieve predetermined purposes.

But during the eighteenth century, traditional theism often gave way to *deism:* the belief that God started the universe and left it to run by itself. In a clockwork world, God's role was restricted to that of clockmaker. The argument from design (natural theology) was initially advanced to support revealed theology, but before long it was taken as a substitute for the latter. What started as reinterpretation within a Christian community rooted in scripture and personal religious experience often ended with a distant and impersonal God. Especially among the Enlightenment philosophers in France, a stronger hostility to the church and all forms of religion developed. A militant atheism was advanced in the name of science, accompanied by great confidence in the power of reason and human progress.

Newton's laws of motion and gravity seemed to govern all objects, from the smallest particle to the most distant planet. The concepts of Newtonian physics were spectacularly successful in explaining a wide range of phenomena, and it is understandable that it was assumed that they could explain *all* events. They were extended to form an all-encompassing metaphysics of *materialism* defended by some of the philosophers of the Enlightenment. *Determinism* was most explicitly defended by Laplace, who claimed that if we knew the position and velocity of every particle in the universe, we could calculate all future events. His claim is *reductionist* in assuming that the behavior of all entities is determined by the behavior of their smallest components.

When Napoleon said to him, "M. Laplace, they tell me that you have written this large book on the system of the universe, and have never mentioned its Creator," Laplace gave his famous reply: "I have no need of that hypothesis."[8]

2. God and Chance

Of the three interpretations of the uncertainties present in quantum theory outlined earlier in this chapter, the first maintains that the world itself is completely *determined*, even though we have not yet found all its exact laws. Einstein, the most prominent defender of this view, believed that there are hidden variables we have yet to discover. The second view ascribes uncertainty to inescapable experimental or conceptual limitations, and it remains *agnostic* about what is going on in the world apart from us. Bohr has commonly been associated with this view. The third view, which Heisenberg advocated, defends *indeterminacy* in nature. The atomic world contains a range of potentialities. The realization of a particular observation or event within a given probability distribution is entirely a matter of chance.

In itself, chance is *random*, whereas divine action is said to be purposeful and goal-directed. Some writers hold that the presence of chance undermines theistic beliefs and supports a materialistic philosophy. Earlier in the century, Bertrand Russell wrote, "Man is the product of causes which had no prevision of the end they were achieving; his origin, his growth, his hopes and fears, his loves and his beliefs, are but the outcome of accidental collocations of atoms."[9] In the previous chapter we saw that some cosmologists have proposed theories postulating many universes (whether in the form of successive cycles of expansion and contraction, isolated domains, Everett's "many worlds," or multiple quantum vacuum fluctuations). The fundamental constants might vary among these universes, and we just happen by chance to live in one that is just right for the emergence of life and consciousness. These cosmologists believe that our existence is the product of chance and not purpose.

Jacques Monod maintains in *Chance and Necessity* that the presence of chance in nature supports materialism and excludes a theistic interpretation. He writes primarily about evolutionary biology, in which chance and necessity are represented by random mutations and natural selection, but he gives examples from other fields of science. He says that the prevalence of chance shows that this is a purposeless universe. "Man knows that he is alone in the universe's immensity, out of which he emerged only by chance."[10] Chance is "the source of all novelty, all creation." Monod holds that all phenomena can be reduced to the laws of physics and chemistry and the operation of chance. "Anything can be reduced to simple, obvious mechanical interactions. The cell is a machine. The animal is a machine. Man is a machine."[11]

There are two possible theological responses to indeterminacy at the quantum level. The first is to say that the choice among the range of possibilities left open by quantum theory is *not a matter of chance*, but is made by God without violating natural laws and without being scientifically detectable. The laws of nature specify only a range of potentialities, but God determines which one is actually realized. The "hidden variable" is God, not a still undiscovered deeper level of deterministic laws. This position is explored in the last section of this chapter.

The second theological response is to assert that *both law and chance are part of God's design*. The biochemist and theologian Arthur Peacocke gives chance a positive role in the exploration of the potentialities inherent in the created order. This approach is consistent with the idea of divine purpose, though not with the idea of a precise predetermined plan. Through the built-in potentialities of higher levels of organization, God could envision a general direction of evolutionary change but not the exact sequence of events. This proposal is explored in the next chapter (see "God and Continuing Creation"). In Chapter 6 we will meet other writers who suggest that the acknowledgment of law and chance (along with waste, evil, and suffering) in nature should lead us to modify or reject classical ideas of divine omnipotence (see "God's Self-Limitation" and "Process Theology").

INDEPENDENCE

Two ideas taken from interpretations of quantum physics have been used to defend the independence of science and religion. First, an *instrumentalist* account of quantum theories can be combined with an instrumentalist account of religious beliefs to argue that science and religion are differing languages serving unrelated functions in human life. Second, the *complementarity* of wave and particle models in quantum physics has been extended to suggest that science and religion provide complementary models of reality that are independent and not in conflict. Proponents of the Independence thesis say that the lessons of quantum theory are epistemological in helping us to recognize the limitations of human knowledge, rather than metaphysical in telling us about the character of reality.

I. Instrumentalist Views of Quantum Theory

Three rival interpretations of the status of theories in quantum physics have been proposed. Each has been used to support a similar interpretation of the status of religious beliefs.

1. *Classical Realism.* Newton and almost all physicists through the nineteenth century said that theories are descriptions of nature as it is in itself, apart from the observer. They saw space, time, and mass as properties of events and objects in themselves. To the classical realist, conceptual models are replicas of the world that enable the scientist to visualize the actual structure of the world. Einstein continued this tradition, insisting that a full description of an atomic system requires specifying the classical spatiotemporal variables that define its state objectively and unambiguously. He held that since quantum theory does not do this, it is incomplete and will eventually be superseded by a theory that fulfills classical expectations.

2. *Instrumentalism.* According to instrumentalism, theories are convenient human constructs, calculating devices for correlating observations and making predictions. Theories are also practical tools for achieving technical control. They are to be judged by their usefulness in fulfilling these goals, not by their correspondence to reality (which is inaccessible to us). Models are imaginative fictions used temporarily to construct theories, after which they can be discarded; they are not literal representations of the world. We cannot say anything about the atom between our observations, though we can use the quantum equations to calculate the probability

that a particular observation will occur in a particular experiment. Theories and models are useful intellectual and practical tools, but they do not tell us anything about the world in itself. This view is known as "the Copenhagen interpretation" because it has often been ascribed to the Dane, Bohr.

3. *Critical Realism*. Critical realists are intermediate between classical realists and instrumentalists. They view theories as partial representations of limited aspects of the world as it interacts with us. Theories, they say, allow us to correlate diverse aspects of the world manifest in differing experimental situations. To the critical realist, models are abstract and selective but indispensable attempts to imagine the structures of the world that give rise to these interactions. The goal of science, in this view, is understanding, not control. The corroboration of predictions is one test for valid understanding (along with coherence and scope), but prediction is not in itself a goal of science. I have defended critical realism in my own writings.[12]

It is often assumed that Bohr must have been an instrumentalist because he rejected classical realism in his protracted debate with Einstein. But actually, he seems to have been closer to critical realism than to instrumentalism. He said that classical concepts cannot be used unambiguously to describe independently existing atomic systems. Classical concepts can be used only to describe observable phenomena in particular experimental situations. We cannot visualize the world as it is in itself, apart from our interaction with it. Henry Folse's study of Bohr concludes: "He discarded the classical framework and kept a realistic understanding of the scientific description of nature. What he rejects is not realism, but the classical version of it."[13] Bohr insisted that we have to abandon the sharp separation of *the observer* and *the observed* that was assumed in classical physics. The limitations of our knowledge are both experimental and conceptual. But Bohr did presuppose the reality of the atomic system that is interacting with the observing system.

Even if it is debatable whether Bohr should be called an instrumentalist (passages in his writing can be cited on both sides), subsequent interpretations of quantum theory have often been *instrumentalist*. The fact that classical variables cannot be used to describe the atomic world is taken as a fundamental conceptual limitation in human knowledge. Some theologians insist that similar conceptual limitations are inescapable in talking about God. Neither science nor religion can say very much about reality in itself, apart from our involvement in it. Instrumentalists are supported by the linguistic analysts who insist that different types of language serve very different functions in human life; science and religion are said to be independent and unrelated enterprises. They serve useful but differing functions, even though neither one provides knowledge of reality in itself.

2. The Lesson of Complementarity

The Complementarity Principle developed by Bohr within quantum physics has been extended and applied to the relation between science and religion. We have seen that in some experiments an electron or a light quantum behaves like a *particle*, while in other experiments it behaves like a *wave*. Bohr emphasized that we must always talk about an atomic system in relation to an *experimental system;* we can never talk about it in isolation, in itself. No sharp line can be drawn between the process of observation and what is observed. We are actors and not merely spectators, and we choose the experimental tools we will employ. Bohr held that it is the interactive process of observation, not the mind or consciousness of the observer, that must be taken into account.

But Bohr also stressed *the conceptual limitations* of human understanding. He shared Immanuel Kant's skepticism about the possibility of knowing the world in itself. If we try to force nature into certain conceptual molds, we preclude the full use of other concepts. We must choose either causal *or* spatiotemporal descriptions, either wave *or* particle models, either accurate knowledge of position *or* momentum. The more one set of concepts is used, the less the complementary set can be applied simultaneously. This reciprocal limitation occurs because the atomic world cannot be described in terms of the concepts of classical physics and observable phenomena.[14]

Bohr himself proposed that the idea of complementarity could be extended to other phenomena susceptible to analysis by two kinds of models: mechanistic and organic models in biology, behavioristic and introspective models in psychology, models of free will and determinism in philosophy, and models of divine justice and divine love in theology. Some authors go further and speak of *the complementarity of science and religion*. Thus C. A. Coulson, after explaining the wave/particle duality and Bohr's generalization of it, calls science and religion "complementary accounts of one reality." This would allow the two disciplines to be relatively independent (though if they are accounts of "one reality" they cannot be completely independent).[15] These ideas are discussed further in "Body and Soul as Complementary Perspectives" (see Chapter 5). Claims that science and religion are complementary languages are discussed further in Chapter 6.

I would argue that the term *complementarity* can be extended from physics to other fields, but only with some caution.[16] First, models should be called *complementary* only if they refer to *the same entity* and are of *the same logical type*. Wave and particle are models of a single entity (e.g., an electron); they are on the same logical level and had previously been employed in the same discipline. These conditions do not apply to science and religion, which are practiced in differing situations and serve differing functions in human life. For these reasons I speak of science and religion as alternative languages, reserving the label *complementary* for models of the same logical type within a given language (such as personal and impersonal models of God).

Second, one should make clear that the use of the term *complementarity* outside physics is *analogical* and not *inferential*. There must be evidence of the value of two alternative models or sets of constructs in the other field. It cannot be assumed that methods found useful in physics will be fruitful in other disciplines.

Third, complementarity provides *no justification for an uncritical acceptance of dichotomies*. It cannot be used to avoid dealing with inconsistencies or to veto the search for unity. The paradoxical element in the wave/particle duality should not be overemphasized. We do not say that an electron is both a wave and a particle, but only that it exhibits wavelike and particle-like behavior. Moreover, we do have a unified mathematical formalism that provides for at least probabilistic predictions. We cannot rule out the search for a new unifying model, even though previous attempts have not yielded theories in better agreement with the data than quantum theory. Coherence remains an important ideal in all reflective inquiry, even if it is qualified by acknowledgment of the limitations of human language and thought.

To summarize this section, we can say that both the instrumentalist interpretation of quantum theory and the Complementarity Principle assert that classical realism is untenable. Some people combine instrumentalism in science with an instrumentalist understanding of religion to support the Independence thesis. I have suggested, however, that critical realism offers a path between classical realism and instrumentalism. It recognizes the limitations of concepts and models in both science and religion but insists that they refer to the real world, though always selectively and inadequately. Models are symbolic representations of a reality that cannot be directly observed or visualized by any one analogy drawn from everyday life. If science and religion refer to a common world, as critical realism holds, there can be significant opportunities for dialogue.

DIALOGUE

Some interpretations of quantum theory go beyond instrumentalism in making metaphysical claims about the world that seem to be relevant to religion. Here quantum physics is said to offer conceptual parallels to ideas in religion. We will examine the role of the observer in quantum physics and the holistic character of quantum systems.

I. The Role of the Observer

The long tradition of philosophical idealism, the belief that reality is essentially *mental* in character, started long before the rise of modern science. The Pythagoreans in ancient Greece held that mathematical relationships are the underlying reality of nature. The Platonists took nature to be an imperfect reflection of another realm of perfect eternal forms. In the seventeenth century, Johannes Kepler said that "geometrical perfection" is the reason that the planets follow exact elliptical orbits, because "God ever geometrizes."

New versions of *philosophical idealism* have claimed support from modern physics. Writing in the 1930s, James Jeans said, "The universe begins to look more like a great thought than like a great machine. Mind no longer appears as an accidental intruder in the realm of matter."[17] Arthur Eddington assigned the determinative influence in all knowledge to the human mind. He pictures us following footsteps in the sand, only to discover that the tracks are our own. We impose our own patterns of law so that "the mind may be regarded as regaining from Nature that which the mind has put into Nature."[18]

In quantum physics, the connection between theory and experiment is very indirect. Instrumentalists stress the experimental side, arguing that theories are only useful fictions for correlating observations. But other scientists, focusing on the theoretical concepts, which are abstract and mathematical, find encouragement for idealistic interpretations.

One major problem arises from *the act of measurement*, in which the multiple potentialities of an atomic system become one actuality. Physicists have been puzzled by the sharp discontinuity that occurs when the wave function (the "superposition of states" representing alternative outcomes) collapses to the one value that is observed. Along the route between the microsystem and the human observer, where does the initially indeterminate result get fixed? Eugene Wigner holds that quantum results are fixed only when they enter somebody's *consciousness:* "It is not possible to formulate the laws in a fully consistent way without reference to consciousness."[19] He maintains that the distinctive feature of human consciousness that causes the wave function to collapse is introspection or self-reference; consciousness gives an account of its own state, cutting the chain of statistical coordinations.

John Wheeler asserts that ours is *an observer-created universe*. The collapse of the wave function is the product of intersubjective agreement in which the key feature is not consciousness but communication. He argues that the past has no existence until it is recorded in the present. He tells the story of a conversation between three baseball umpires. One says, "I calls 'em as I sees 'em." The second claims, "I calls 'em as they really are." The third replies, "They ain't nothin' until I calls 'em." Wheeler says that as observers of the Big Bang and the early universe, we have helped to create those events. Before there were observers, atoms were only partially individuated; they had enough reality to enter chemical reactions but were not fully real until they were later observed. He grants that it seems an anomaly that the present could influence the past, but he says that, in the quantum world of indeterminacy, ideas of before and after are meaningless. The past has no meaning unless it exists as a record in the present. So human beings are central in a participatory and observer-dependent universe.[20]

I do not find these interpretations of quantum physics convincing. Surely it is not mind as such that affects observations, but the process of *interaction* between the detection apparatus and the microsystem. The experimental results might be automatically recorded on film or on a computer printout that no one looks at for a year. How could looking at the film or printout alter the outcome of an experiment that has been recorded for a year? The Wheeler view seems very strange, for observers of the Big Bang are themselves products of the evolution of the cosmos, which included billions of years when there was neither consciousness nor observer. Atoms that affect subsequent evolutionary events must surely be considered fully real. In all these cases, I would say, the lesson to be learned is that phenomena in the world are interdependent and interconnected, not that they are mental in character or intrinsically dependent on the human mind.

Moreover, some ingenious experiments in the 1990s have made it possible to study the *decoherence* of a quantum wave function as it interacts with a wider environment. A stream of sodium atoms or beryllium ions has been probed by laser pulses along its path to investigate the transition from quantum to classical behavior. The coherence of a quantum state is lost when information about it is available through interaction with the laser pulses, which can be considered a form of "measurement." The transfer of information, not consciousness, is the essential feature of the "collapse of the wave function" during an observation.[21]

But contemporary physics does have an epistemological lesson about *the participation of the observer.* In quantum physics the observer participates through the interactive character of observation processes. In relativity, temporal and spatial properties vary with the frame of reference of the observer; these properties are now understood to be *relationships* rather than intrinsic features of separate objects in themselves. In religion, too, knowledge is possible only by participation, though of course the forms of participation differ from those in science. We can ask how God is related to us, but we can say little about the intrinsic nature of God.

It is sometimes said that quantum physics is *less materialistic* than Newtonian or nineteenth-century physics. Probability waves may seem less substantial than billiard-ball atoms, and matter that converts to radiant energy may appear immaterial. But the new atom is no more spiritual or mental than the old, and it is still detected through physical interactions. If science is indeed selective and its concepts are limited, as I have argued, it would be as questionable to build a metaphysics of idealism on modern physics as it was to build a metaphysics of materialism on classical physics.

2. Holism in the Quantum World

Beyond the challenges to determinism and realism, quantum physics also challenges the *reductionism* of classical physics. What were once thought to be "elementary particles" seem to be temporary manifestations of shifting patterns of waves that combine at one point, dissolve again, and recombine elsewhere. A particle begins to look more like a local outcropping of a continuous substratum of vibratory energy. A force between two particles (protons, for example) can be thought of as arising from a field, or from a rapid exchange of other kinds of particles (mesons, in this case). A bound electron in an atom has to be considered as a state of *the whole atom* rather than as a separate entity. As more complex systems are built up, new properties appear that were not foreshadowed in the parts alone. New wholes have distinctive principles of organization as systems, and therefore exhibit properties and activities not found in their components.

Consider the helium atom. In the prequantum model it was pictured as a nucleus around which circled two separate identifiable electrons. The atom's parts were clearly distinguishable, and the laws of its behavior were thought to be derivable from analysis of the behavior of these components. But in quantum theory the helium atom is a total pattern with *no distinguishable parts*. Its wave function is not at all the sum of two separate single-electron wave functions. The electrons have lost their individuality; we do not have electron A and electron B, but simply a two-electron pattern in which all separate identity is lost. (In the statistics of classical physics, an atom with electron A in an excited energy state and electron B in a normal state counts as a different configuration from the atom with A and B interchanged, but in quantum theory it does not.)

In the case of larger atoms with additional electrons, we find that their configurations are governed by the *Pauli exclusion principle*, a law concerning the total atom that cannot conceivably be derived from laws concerning individual electrons. This principle asserts that in a given atom no two electrons can be in identical states (with the same quantum numbers specifying energy, angular momentum, and spin). To this remarkable and far-reaching principle can be attributed the periodic table and the chemical properties of the elements. When another electron is added to a given atom, it must assume a state different from all electrons already present. If one used classical reasoning, one would have to assume that the new electron is somehow influenced by all the other electrons; but this "exclusion" does not resemble any imaginable set of forces or fields. In quantum reasoning any attempt to describe the behavior of the constituent electrons is simply abandoned; the properties of the atom as a whole are analyzed by new laws unrelated to those governing its separate "parts," which have now lost their identity. An orbital electron is a state of the system, not an independent entity.

The energy levels of an array of atoms in the solid state (such as a crystal lattice) are a property of the whole system rather than of its components. Moreover, some of the disorder-order transitions and the so-called cooperative phenomena have proven impossible to analyze through the behavior of the components—for example, the behavior of electrons in a superconductor. There are *system laws* that are not derivable from the laws of the components; distinctive explanatory concepts characterize higher organizational levels. Interpenetrating fields and integrated totalities replace self-contained, externally related particles as fundamental images of nature. The being of any entity is constituted by its relationships and its participation in more inclusive patterns.[22]

An impressive form of holism known as *nonlocality* is shown in a type of experiment proposed by Einstein and more recently by John Bell. In one version, a source emits two photons, A and B, which fly off in opposite directions—left and right, let us say. One detector at the left measures the spin of A. The spin of B is immediately known; it is equal and opposite to that of A, since the system initially had total spin zero. In a 1997 experiment, the left detector was thirty kilometers distant at the far end of an optical fiber. The orientation of this detector was chosen while A was in flight—far too late for a signal to reach B before it arrived at the right detector, assuming that no signal can travel faster than the speed of light, as relativity theory requires.[23]

Quantum theory describes each photon in flight as a mixture ("superposition") of waves representing all possible spin orientations.

Each set of waves collapses to a single value only when a measurement is made. How could the waves representing B know what is happening to the waves representing A? The connection seems to be instantaneous and does not decrease with distance as most physical forces do. However, we could not use such a system to send a message faster than the speed of light (in violation of relativity theory) because the spin of A is unpredictable; it can be recorded but not controlled by the experimenter. Evidently the two particles originating in one event must be described by a single overall wave function, no matter how far apart they are. Polkinghorne concludes, "Quantum states exhibit a surprisingly integrationist view of the relationship of systems which have once interacted with each other, however widely they may subsequently separate."[24]

Jean Staune holds that such scientific knowledge "demonstrates the existence of a level of reality that escapes time, space, energy, and matter, yet still has a causal effect on our material level of reality." He maintains that this is significant not only in refuting materialism but also in giving us a way to speak about God. "Science has suggested through quantum physics that it alone cannot provide a complete picture of reality. It has provided the basis for a credible way to understand the existence of God, because the world no longer limits itself to our level of reality."[25]

I believe that the various forms of quantum holism that I have described are indeed significant as a critique of reductionism. However, I am cautious about relying too heavily on these two-particle experiments alone, because the relation of quantum theory to relativity is so complex and controversial.[26] The theme of holism is carried further in the next chapter (see "Complexity and Self-Organization" and "A Hierarchy of Levels"). Its theological implications are discussed in Chapter 6 (see "God as Top-Down Cause").

INTEGRATION

Proponents of Integration claim a closer relation between scientific theories and particular religious beliefs than that set forth by advocates of Dialogue, though there is no sharp line separating the groups. Two versions are explored here, one drawing from quantum holism and the other from quantum indeterminacy.

1. Eastern Mysticism and Quantum Holism

Several authors have offered a systematic integration of contemporary physics and Eastern mysticism.[27] The most influential and widely read volume is *The Tao of Physics* by the physicist Fritjof Capra, which starts by setting forth epistemological similarities. According to Capra, both physics and Asian religions recognize *the limitations of human thought and language*. Paradoxes in physics, such as the wave/particle duality, are reminiscent of the yin/yang polarity in Chinese Taoism, which portrays the unity of apparent opposites; Bohr himself put the yin/yang symbol at the center of his coat of arms. Zen Buddhism asks us to meditate on koans, the paradoxical sayings to which there is no rational solution. Capra also says that mind plays an essential role in the construction of reality: "Ultimately, the structures and phenomena we observe in nature are nothing but the creations of our measuring and categorizing minds."[28] He cites Wigner's assertion that quantum variables have no definite values until the intervention of human consciousness.

But Capra goes further in holding that both physics and Eastern mysticism make similar metaphysical claims about *the wholeness of reality*. Quantum physics points to the unity and interconnectedness of all events. Particles are local disturbances in interpenetrating fields. In relativity, space and time form a unified whole, and matter-energy is identified with the curvature of space. Eastern thought also accepts the unity of all things and speaks of the experience of undifferentiated oneness encountered in the depth of meditation. There is one ultimate reality, referred to as Brahma in India and the Tao in China, with which the individual is merged. The new physics says that the observer and the observed are inseparable, much as the mystic tradition envisages the union of subject and object.

Next, Capra holds that both physics and Eastern thought treat the world as *a dynamic and ever-changing system*. Particles are patterns of vibration that are continually being created and destroyed. Matter appears as energy, and vice versa. Hinduism and Buddhism hold that life is transitory; all existence is impermanent and in ceaseless motion. The dance of Shiva is an image of the cosmic dance of form and energy. But in both modern physics and Asian religions there is an underlying *timeless realm*. Capra maintains that in relativity theory spacetime is timeless, like the eternal now of mystical experience.

I think Capra has overstressed the similarities and virtually ignored the differences between the two disciplines. Often he finds a similarity by comparing particular terms or concepts, abstracted from wider contexts that are radically different. For example, Asian traditions speak of undifferentiated unity. But the wholeness and unity that physics expresses is highly differentiated and structured, subject to strict constraints, symmetry principles, and conservation laws. Space, time, matter, and energy are all unified in relativity, but there are exact transformation rules. The mystic's structureless unity, in which all distinctions are obliterated, seems very different from the organized interaction and cooperative behavior of higher-level wholes, seen already in physics but much more evident in biology. If reductionists see only the parts, Capra gives one-sided attention to wholes.

I believe that the relation between *time* and *timelessness* is also significantly different in physics and in mysticism. Physics deals with the realm of temporal change. I agree with Capra that in the atomic world there is impermanence and an ever changing flux of events. But I do not agree that spacetime is a static and timeless block. I have argued that the unity of space and time in relativity points to the temporalization of space rather than the spatialization of time.[29] On the other hand, for much of Eastern mysticism, especially the Advaita tradition in Hinduism, the temporal world is illusory and ultimate reality is timeless. Beneath the surface flux of maya (illusion) is the unchanging center, which alone is truly real, even though the world exhibits regular patterns to which a qualified reality can be ascribed. In Buddhism, timelessness refers to the realization of our unity with all things, which releases us from bondage to time and the threat of impermanence and suffering. Meditative disciplines do bring the experience of a sense of timelessness, though I suggest that this may be partly the product of absorptive attention that stops the flow of thought.

Capra says little about *the difference in the goals* of physics and mysticism or the distinctive functions of their languages. The goal of meditation is not primarily a new conceptual system, but the transformation of personal existence, a new state of consciousness and being, an experience of enlightenment. Mysticism is a way of life and only secondarily a set of metaphysical beliefs.

The physicist David Bohm is more cautious in delineating similarities between physics and Eastern mysticism. He proposes that mind and matter are two different projections of an underlying *implicate order;* they are two related expressions of a single deeper reality. Bohm also finds in Eastern religions a recognition of the basic unity of all things; in meditation there is a direct experience of undivided wholeness. Fragmentation and egocentricity can be overcome in the absorption of the self in the undifferentiated and timeless whole.[30] Here is an ultimate monism that contrasts with the greater pluralism of Western religions. For Bohm, the answer to the fragmentation of personal life is the dissolution of the separate self, whereas Christian thought seeks the healing of brokenness by the restoration of relationships with God and neighbor.

Richard Jones gives a detailed comparison of themes in the new physics, Advaita Hinduism, and Theravada Buddhism; he emphasizes the *differences* among them.[31] He subscribes basically to what I have called the Independence thesis: science and mysticism are distinct and separate, but both have cognitive value. Science has authority concerning objective structures and regularities in the realm of becoming and change, while mysticism is an experience of the unstructured, non-objectifiable reality beneath the surface multiplicity. For the most part, their claims are incommensurable, and no integration is possible, for they refer to different realms. Science deals objectively with differentiated lawful structures, while the mystic encounters the undifferentiated wholeness of the underlying reality in the experience of meditation. Jones is critical of the vague parallels that Capra draws and the use of phrases abstracted from their contexts. Perhaps Jones draws too sharp a line between science and religion, but he reminds us of the dangers of basing Integration on similarities without considering differences.

2. God and Quantum Indeterminacy

Some authors have suggested that atomic indeterminacies are the domain in which God *providentially controls the world.* In the 1950s, William Pollard, a physicist and priest, proposed that such divine action would violate no natural laws and would not be scientifically detectable. God, he suggests, determines which actual value is realized within the range of a probability distribution. The scientist finds no natural cause for the selection among quantum alternatives; chance, after all, is not a *cause.* The believer, on the other hand, may view the selection as God's doing. God would influence events without acting as a physical force. Since an electron in a superposition of states does not have a definite position, no force would be required for God to actualize one among the set of alternative potentialities. By a coordinated guidance of many atoms, God could providentially govern all events. According to Pollard, it is God, not the human observer, who collapses the wave function to a single value.[32]

Indeterminacy at the quantum level at first seems to be irrelevant to phenomena *at the level of a living cell* containing millions of atoms, among which statistical fluctuations tend to average out. Quantum equations give exact predictions for large ensembles, though not for individual events. Moreover, atoms and molecules have an inherent stability against small perturbations, since at least a quantum of energy is required to change their states. However, in many biological systems individual microevents can have large-scale consequences. One mutation in a single component of a genetic sequence can change evolutionary history. In the nervous system and the brain, a microevent can trigger the firing of a neuron whose effects are amplified by the neural network. By controlling quantum events, then, God could affect events in evolutionary and human history.

Pollard's proposal is consistent with current theories in physics. God would be the ultimate nonlocal "hidden variable." But I have three objections to his ideas: (1) Pollard asserts divine sovereignty as *total control* over all events, and he defends predestination. This seems to me incompatible with human freedom and the reality of evil. It also denies the reality of chance, which he thinks is only a reflection of human ignorance of the true divine cause. (2) For Pollard, God's will is achieved through the *unlawful* rather than the *lawful* aspects of nature. This may be a needed corrective to deism's opposite emphasis, but it seems equally one-sided. (3) There is an *implicit reductionism* in assuming that God acts only at the lowest level, that of the atomic components. Do we not want to allow also for God's influence on higher levels, "from the top down" rather than "from the bottom up"? Is God not related directly to the integrated human self, for example, and not just to the atomic events in the brain?

The physicist and theologian Robert Russell is among those who hold that God influences *only certain quantum events* and also acts at higher levels as a top-down cause on events at lower levels. This would avoid the objections raised against Pollard's scheme, and it would allow for chance, law, and God's action in the quantum world. If an event is not completely determined by lawful relationships, its final determination might in some cases be made directly by God, who would actualize one rather than another of the system's potentialities. Since this would not violate the statistical laws of quantum theory, it would not be detectable by the scientist. The proposal that God acts in quantum indeterminacies is not intended as an argument for the existence of God in the tradition of natural theology, since uncertainties might still be attributable to human ignorance or to chance. The proposal is offered rather as *a theology of nature*—that is, a way in which the God in whom we believe on other grounds might be conceived to act in ways consistent with scientific theories.[33] We will examine this proposal further in Chapter 6 (see "God as Determiner of Indeterminacies").

Let me then summarize my own conclusions.

1. *Law and Chance.* The determinism of natural laws in classical physics was often seen as a challenge to traditional ideas of God's action in the world. In quantum physics the randomness of observable events within a range of probable values might also seem incompatible with the concept of divine sovereignty. But some authors have suggested ways in which traditional ideas of God's role can be reformulated to acknowledge both law and chance, as explored further in Chapters 4 and 6.

2. *Critical Realism.* Quantum physics shows the deficiencies of classical realism, but critical realism allows us to acknowledge the limitations of human knowledge without adopting the agnosticism about events in the world typical of instrumentalism. Models and theories are not literal descriptions but refer selectively and inadequately to limited aspects of reality.

3. *Indeterminacy.* In accordance with critical realism, the uncertainty in quantum physics is understood as a reflection of indeterminacy in nature and not simply as a product of human ignorance of events that are themselves determined. In this view, alternative potentialities are held to be present in the world.

4. *Complementarity.* The need to use pairs of models and concepts that cannot be unified coherently (such as *wave* and *particle*) reflects the limitations of human thought in a domain far from everyday experience. But complementarity does not rule out the search for unity, nor does it support the claim that scientific and religious assertions are totally unrelated and independent.

5. *The Involvement of the Observer.* The observer acting through the experimental apparatus influences observations of quantum events. I have interpreted this as an example of holism (the need to take the whole physical system into account) rather than as evidence of the influence of the observer's mind or the pervasiveness of consciousness.

6. *Holism.* The holistic character of quantum theory is evident in the use of wave functions for a whole atomic configuration that cannot be analyzed as the sum of its separate parts. It is also seen in the correlation of distant events in Bell's theorem experiments. Such holism in physics and other fields supports the rejection of reductionism. It is consistent with a multilevel view of reality and the emergence of new kinds of events at higher levels of organization.

7. *Eastern Mysticism.* The monism of Eastern traditions seems to me to be a holism that is too extreme and that diminishes the reality and relationality of individual beings. Its emphasis on timelessness is compatible with some versions of Christian thought, but not with the process view of God's relation to time (which I will explore later).

8. *God as Determiner of Indeterminacies.* The traditional view of divine omnipotence can still be maintained without violating the laws of physics if God is the determiner of indeterminacies at the quantum level. But I will later conclude that the ideas of divine self-limitation and process theology are more consistent with both scientific evidence and central Christian beliefs.

Chapter Four

Evolution and Continuing Creation

Serving as naturalist on the HMS *Beagle* during a five-year voyage around the world, Charles Darwin observed many variations within individual species in different locations. Six years later, reading Malthus's essay on human populations competing for limited resources, he found the clue for a theory to interpret the voluminous data he had collected on the voyage. Darwin had noted the gradual changes in successive generations of animals when human breeders select for a particular trait. In every population there are small random variations that can be inherited. In the struggle for survival in nature, he argued, some variations confer a slight competitive advantage, leading over a period of many generations to the "natural selection" of heritable characteristics that contribute to survival. In 1859, in *On the Origin of Species*, he elaborated the thesis that new species have come into existence by variation and selection over a long period of time.

Work in *population genetics* in the twentieth century greatly advanced our understanding of the inheritance of variations, as Mendel's laws of heredity were studied in plant, insect, and animal populations. It was also found that an occasional individual had a characteristic, such as eye color, that differed markedly from those of the rest of the population. The frequency of such abrupt changes (mutations) could be increased by exposure to X-rays and certain chemicals. Mutations and the combination of units of heredity (genes) from two parents were seen to be the main sources of variation, and both were evidently random processes unrelated to the needs of the organism. Genetics and evolutionary theory were brought together in a systematic neo-Darwinism to which Julian Huxley in 1942 gave the name "the Modern Synthesis." Among its exponents were Ernst Mayr, Theodosius Dobzhansky, and Gaylord Simpson.[1]

In 1953, James Watson and Francis Crick discovered the structure of *the DNA molecule* and identified it with the genes that had been postulated in population genetics. A particular sequence of three bases within a strand of DNA corresponds to a particular amino acid. The order of the triplets of bases in the DNA determines the order in which the amino acids are assembled into the protein chains that form the cells of all living organisms. In DNA, an "alphabet" of just four "letters" (the bases), grouped in three-letter "words" (triplets, each specifying one of the amino acids), is arranged in "sentences" (specifying particular proteins). Thousands of sentences of varying length and word order can be made from the twenty basic words, so there are thousands of possible proteins. Long strands of DNA, made of exactly the same four bases in various sequences, constitute the genes of all organisms, from microbes to human beings. In all known organisms, the same genetic code is used to translate from DNA to amino acid, which seems to indicate a common origin for all living things.

Recent techniques for *comparing the molecular structure* of similar proteins in various living species allow us to estimate the time since their lineages diverged. For example, the enzyme cytochrome-C in human beings consists of a sequence of 104 amino acids. In the comparable sequence in rhesus monkeys, only one of these amino acids is different; horses have twelve that differ and fish have twenty-two, indicating increasingly distant kinship. The evolutionary history established by this biochemical method agrees well with evidence derived from two completely different scientific fields: the study of fossil records by paleontologists and the comparison of the anatomy of living species by taxonomists.[2]

The study of fossil forms has also raised important questions about evolutionary history. Neo-Darwinism since 1940 has shared

Darwin's assumption that long-term evolutionary changes are the result of the gradual accumulation of many small changes. But in the 1970s Stephen Jay Gould and Niles Eldredge advocated *punctuated equilibrium*, the theory that there have been long periods of stability interrupted by brief periods of rapid change. They pointed to fossil records that show millions of years with very little change, interspersed with bursts of rapid speciation in relatively short periods—especially in the early Cambrian period, when all the known evolutionary groups (phyla) and basic body plans appeared in a very short period. They postulated that alterations in developmental sequences produced major structural changes. Their view is holistic in claiming that most traits are the product of many genes. They directed attention to the whole set of genes and the role of regulatory programs in embryonic development, rather than to mutations in single genes. The directions of change, they said, are determined by the possibilities of developmental reorganization as well as by selective forces acting on adult organisms. Moreover, the extinction of species is sometimes the product not of gradual competitive forces but of sudden contingent events, such as the impact of comets.[3]

Defenders of neo-Darwinism replied that their theory was more varied and flexible than Gould and his supporters had acknowledged. The rarity of transitional forms among fossils may be a result of the incompleteness of the fossil record. Changes that appear rapid on the scale of geological time (over a period of fifty thousand years, let us say) can encompass many generations. Thus Ledyard Stebbins and Francisco Ayala said that many of Gould's ideas could be included in an expanded version of the neo-Darwinian synthesis.[4]

Neo-Darwinism viewed evolutionary change as the product of random variations that were then selected by the environment. But some biologists have noted that the *internal drives of organisms* can initiate evolutionary changes. The environment selects individuals, but individuals also select environments, and in a new niche a different set of genes may contribute to survival. A few pioneering fish ventured onto land and were the ancestors of amphibians and mammals; some adventurous mammals later returned to the water and were the ancestors of dolphins and whales. In each case organisms themselves took new initiatives; genetic and then anatomic changes followed from their actions (the so-called Baldwin effect).[5]

In such cases, random mutations in genes did not start the chain of events that led to change; rather, they served to perpetuate changes first introduced by *the actions of the organisms* themselves. This does not imply that organisms were *trying* to evolve, only that purposive behavior as well as chance mutations set the direction of evolutionary change. Biologists favoring this view do not adopt the discredited Lamarckian idea that physiological changes acquired during an organism's lifetime can be directly inherited, but they agree with Lamarck that purposive actions can eventually lead to physiological changes. The Baldwin effect can be incorporated in an expanded neo-Darwinism, but it requires revisions of earlier assumptions.

In this chapter we will ask how advocates of each of our four basic views of science and religion interpret current evolutionary theory.

CONFLICT

One version of the Conflict thesis is presented by evolutionary materialists such as Richard Dawkins and Daniel Dennett. A very different version is articulated by theistic critics of neo-Darwinism, such as Phillip Johnson and Michael Behe. But the two versions agree that there is a conflict; they both assert that one cannot with intellectual integrity be both a theist and a neo-Darwinist.

I. Evolutionary Materialism

The Blind Watchmaker, by the biologist Richard Dawkins, carries the subtitle *Why the Evidence of Evolution Reveals a Universe without Design.* Much of the book is a clear and forceful presentation of current evolutionary theory and a defense of orthodox neo-Darwinism against its religious critics. For example, some critics have said that the various parts of the eye (retina, diaphragm, lens, etc.) could not be products of separate chance mutations because one part would be useless without all the other coordinated parts. But Dawkins shows that the eye could have arisen from many small improvements. A rudimentary light-sensitive cell or a very simple eye is better than nothing. Image-forming eyes have evolved independently at least forty times among vertebrates, and nine distinct eye structures can be identified (including pinhole, lens, curved reflector, and compound eyes). Moreover, says

Dawkins, many systems in nature are far from perfect in their design. The human eye is "wired backward"; the nerves come out of the retina on the side toward the lens, obstructing the passage of light and requiring a hole in the retina for the nerves to leave the eye (the blind spot). No sensible designer would have made it that way, but evolutionary change always has to start from what is available and improve it, even if the end result is not the most efficient design.[6]

In a more recent book, Dawkins asks how cheetahs (which are well equipped to kill gazelles) and gazelles (which are well designed to avoid being killed by cheetahs) could both have been designed by the same God. But if the quantity maximized in natural processes is *the survival of DNA*, then the characteristics of both cheetahs and gazelles can be explained. As another argument against design Dawkins points to the widespread suffering, pain, and fear in nature and the senseless tragedies that occur in human life:

> In a universe of blind physical forces and genetic replication, some people are going to get hurt, and other people are going to get lucky, and you won't find any rhyme or reason in it, nor any justice. The universe that we observe has precisely the properties we should expect if there is, at bottom, no design, no purpose, no evil, and no good, nothing but blind, pitiless indifference. . . . DNA neither cares nor knows. DNA just is. And we dance to its music.[7]

Dawkins accepts *epistemological reductionism:* "The hierarchical reductionist believes that carburetors are explained in terms of smaller units . . . which are explained in terms of smaller units . . . which are ultimately explained in terms of the smallest of fundamental particles. . . . My task is to explain elephants, and the world of complex things, in terms of the simple things that physicists either understand, or are working on."[8] He holds the view that science is the only acceptable form of explanation; if science does not discover purpose, there *is* no purpose in the universe. This leads him to an *ontological reductionism* or *materialism*, in which matter is the fundamental reality. Dawkins describes a fascinating computer program that displays insect-like geometrical forms (biomorphs) on the screen, modifies them by random variations, and then selects by criteria built into the computer program. He takes this to be a demonstration that evolution can occur without purpose or intelligent design. In reply, one could say that it shows the opposite: that chance and selection are compatible with intelligent design by a purposive agent—namely, the computer programmer.

The philosopher Daniel Dennett has defended a strong neo-Darwinian position, drawing from biology, probability theory, cognitive science, and computer simulations. Evolution, he asserts, is the product of *a mindless, purposeless process*. He vehemently rejects all forms of intelligent design, including Darwin's belief that the laws of evolution rather than individual species were the product of design. In a protracted critique of Gould he insists that mutations and natural selection are the *only* factors responsible for the direction of evolutionary change. Through random mutations a population's genes explore the neighboring portions of "design space" (the set of all possible genetic configurations). Through natural selection, those genes that confer adaptive advantages are passed on with greater frequency. Selection is thus an automatic, impersonal process following an algorithm (a formal rule with simple steps).[9]

The status of *design space* in Dennett's thought is not altogether clear. In some passages he compares it to the Platonic order of eternal forms (though of course for him the forms are not ideas in the mind of God, as they were in classical thought). In other passages he treats design space as the abstract set of all genetic possibilities, with little inherent structure. In any case, the exploration of design space occurs entirely by chance. The subsequent retention of new configurations is the product of contingent environmental conditions and the usefulness of certain general capacities, such as vision, locomotion, and intelligence. Dennett gives this summary:

What is design work? It is the wonderful wedding of chance and necessity, happening in a trillion places at once, at a trillion different levels. And what miracle caused it? None. It just happened to happen, in the fullness of time. You could even say, in a way, that the Tree of Life created itself. Not in a miraculous, instantaneous whoosh, but slowly, slowly, over billions of years.[10]

Even the laws of physics, he suggests, "could themselves be the outcome of a blind, uncaring shuffle through Chaos."[11]

Like Dawkins, Dennett merges evolutionary science and a philosophy of naturalism. He says that acceptance of evolution requires the rejection of theism. We have seen that for materialists chance and necessity in astronomy and quantum physics are taken to be incompatible with purpose and design; here a similar assertion is made about chance and necessity in evolutionary history.

2. Theistic Critics of Neo-Darwinism

At the time of the Scopes trial in 1925, Christian fundamentalists rejected all evolutionary ideas and accepted a literal interpretation of Genesis. In the 1980s the *"creation science" movement* put pressure on state legislatures and local school boards to require biology teachers to give a "balanced treatment" of evolutionary theory and "abrupt appearance theory" as two alternative scientific hypotheses, claiming that the latter could be defended on scientific grounds alone. State and federal courts overruled legislation that required the inclusion of "creation science," arguing that it is not legitimate science and that teaching it violates the separation of church and state (see Chapter 1). But several authors have offered more sophisticated arguments asserting that *"intelligent design"* fits the data better than purposeless chance and natural selection. They claim that neo-Darwinism among biologists is more the product of their naturalistic assumptions than of clear scientific evidence.[12]

Phillip Johnson, a law school professor, accepts *microevolution* (small modifications in existing species) but does not accept *macroevolution* (the formation of new species). He points out that artificial selection by animal breeders produces great diversity but no new species. The beaks of finches on the isolated Galápagos Islands changed when climate conditions changed, but no new species appeared. Artificial selection of fruit flies has produced forms that do not breed with their ancestors, but it is not clear that they should count as a new species.[13]

Johnson is impressed by the paucity of *transitional forms* in the fossil record. He says that *Archaeopteryx*, a birdlike creature with reptilian features, is one of the few plausible candidates for a link between species. He grants that fossil records, especially of soft body parts, are inevitably rather fragmentary. If Gould is correct that speciation occurred rapidly in small, isolated populations, transitional fossils would be rare. But Johnson claims that even then we would expect more transitional forms than have been found. He holds that attempts to reconstruct a family tree from apes to humans are particularly speculative and subjective, influenced as much by philosophical pre-conceptions as by clear evidence.

The sudden appearance in the early Cambrian period (about 570 million years ago) of creatures very unlike anything found in Pre-cambrian strata is particularly puzzling. As many as one hundred new phyla (groups of similar species with a common body plan) appeared in a relatively short interval, though only thirty of these remain today. Thereafter, few new phyla appeared—even when mass extinc-tions (such as those caused by the impact of comets) opened up new ecological niches; instead, extensive diversification occurred in exist-ing phyla. Moreover, many species changed little over periods of mil-lions of years. Paleontologists had often ignored such *stasis* because they were looking for change. Johnson accepts Gould's concept of punctuated equilibrium, though unlike Gould he thinks that the Cambrian explosion was a product of God's intervention through the communication of new genetic information.

Johnson also argues that random mutations could not have pro-duced *the coordinated functioning* of many parts that occurs in com-plex organisms. An effective eye, for example, requires not only the coordination of diverse parts but also the presence of neural and cerebral structures. He replies to Dawkins's speculation that webs on the limbs of small tree-climbing animals would have enabled them to glide, leading eventually to wings for flying. Johnson says that such webs would have hindered climbing and food-grasping long before they could have supported flight. He says that Dar-winists have elevated chance into an ultimate principle that is inherently anti-theistic.

Scientific critics say that Johnson exaggerates the deficiencies of Darwinian theory. Theories applying to the distant past cannot of course be proved with certainty. But broad theories in science are judged in part by their ability to explain a wide range of data of dif-fering kinds. Scientists compare alternative theories and evaluate their fruitfulness in suggesting further research, but Johnson offers no empirically testable alternative theory.[14] *Theological critics* point out that Johnson assumes that theism requires belief that God inter-venes in gaps in the scientific account. They suggest that he has not adequately distinguished scientific theories themselves from the philosophical positions of their atheistic interpreters. So he has ended by agreeing with the exponents of evolutionary materialism that one cannot believe in both God and neo-Darwinism.

The biochemist Michael Behe argues that the "irreducible complexity" of biochemical systems shows that they cannot be the product of gradual evolution. He traces, for example, the long chains or cascades of molecular reactions that occur in the eye and in the immune system. He claims that such complex interlocking systems could not have had simpler functional precursors because they would not have worked if even a single step or component had been missing. He offers the analogy of a mousetrap with five essential parts. If one part, such as the spring, were missing, the trap would not work at all. It is an all-or-nothing system that must have been designed all at once, not by stages. "Since natural selection can only choose systems that are already working, then if a biological system cannot be produced gradually it would have to arise as an integrated unit, in one fell swoop, for natural selection to have anything to act on."[15]

Behe suggests that *the information for designed systems* might have been introduced in the earliest single cells and remained dormant for billions of years, or it might have been added later to produce a complex biochemical system. He writes:

> Work could be undertaken to determine whether information for designed systems could lie dormant for long periods of time or whether the information would have to be added close to the time when the system became operational. Since the simplest possible design scenario posits a single cell—formed billions of years ago—that already contained all information to produce descendant organisms, other studies could test this scenario by calculating how much DNA would be required to code the information (keeping in mind that much of the information might be implicit).[16]

Both of Behe's scenarios seem to assume a predetermined divine plan of intelligent design rather than a dynamic and open-ended process in which interaction with a changing environment plays a crucial role.

Most *reviews by scientists* have been critical of Behe. They point out that many evolutionary changes can be understood as improvisations using components already available rather than as total systems designed from scratch. Sometimes a component such as an enzyme can serve more than one function and can be co-opted for new functions under altered circumstances. Other critics point to the great progress that has been made in understanding genetic information feedback systems in biological development and regulation today.[17] *Theological critics* have said that by ruling out evolutionary explanations Behe has given a new version of the God-of-the-gaps. He has taken gaps in our knowledge to be gaps in the processes of nature in which God intervenes discontinuously. Edward Davis calls this "a highly sophisticated form of special creationism," which he contrasts with the idea of God as primary cause operating through a seamless web of secondary causes.[18]

INDEPENDENCE

No conflict between science and religion can occur if they are completely independent enterprises that differ in their methods, their domains, and their functions in human life. We will first consider the contrasting domains and methods of biology and theology. We will then look at the assertion that the questions that lead theologians to talk about God as primary cause are very different from the questions about secondary causes within the realm of nature that scientists investigate.

I. Contrasting Domains and Methods

In responding to the challenge of creationism, the National Academy of Sciences published a pamphlet in 1984 that tried to defuse the conflict over biology courses in the public schools by insisting that religion and science have *nothing to do with each other.* "Religion and science are separate and mutually exclusive realms of human thought whose presentation in the same context leads to misunderstanding of both scientific theory and religious belief."[19] Such a "two realms" position can be used to protect high school science teachers from interference by religious groups.

In his recent book *Rocks of Ages: Science and Religion in the Fullness of Life*, Stephen Jay Gould presents science and religion as independent realms. He says that he grew up in a Jewish family but considers himself an agnostic who has great respect for religion. His basic principle is non-overlapping magesteria (NOMA). A magesterium is a domain of teaching authority. "The magesterium of science covers the empirical realm: what is the universe made of (fact) and why does it work this way (theory). The magesterium of religion extends over questions of ultimate meaning and moral value."[20] Each domain has its own distinctive questions, rules, and criteria of judgment. Gould discusses various historical cases in which religious leaders took dogmatic stands outside their field of competence, from Galileo to the Arkansas creationism trial. But he is equally critical of scientists who try to derive philosophical, theological, or ethical conclusions from science. He rejects the attempts of sociobiologists to ground moral judgments in the adaptive value of moral behavior in evolutionary history. He points out that Darwinism has been misused to defend war, colonialism, ruthless economic competition, and eugenics.

Gould is critical of any natural theology that expects science to yield religious conclusions. He dismisses the fine-tuning of the early universe (the Anthropic Principle) and all arguments from design in the biological world. He says that Darwin's belief in the design of evolutionary laws rather than particular species was only his "personal preference," not a "false fusion" of two domains that cannot be brought together. Yet when Gould gives his own views in the final chapter he goes beyond strict adherence to NOMA. He makes statements that appear to be supported by science, but I suggest that they should be viewed as naturalistic philosophical interpretations. He says, for example, that in the face of our "cosmic insignificance" and "the sublime indifference of nature" we can find meaning only within ourselves:

> [Mankind is] a wildly improbable evolutionary event, and not the nub of universal purpose. . . . We are the offspring of history, and must establish our own paths in this most diverse and interesting of conceivable universes—one indifferent to our suffering, and therefore offering us maximal freedom to thrive, or to fail, in our own chosen way.[21]

The independence of science and religion can also be defended from the theological side. *Neo-orthodoxy* has no difficulty accepting the findings of evolutionary biology, because it holds that God acts in human history, primarily in the person of Christ, rather than in the natural world. It finds the argument from design and all forms of natural theology suspect for relying on human reason rather than on divine revelation. According to neo-orthodoxy, the doctrine of creation is not a theory about beginnings or about subsequent natural processes; it is an affirmation of dependence on God and the essential goodness and orderliness of the world. I have indicated my sympathy with some aspects of neo-orthodoxy, particularly its conviction that scripture should be taken seriously but not literally and its assertion that the doctrine of creation *ex nihilo* is an affirmation of an enduring relationship between God and the world and not a description of events in the distant past. However, its strong emphasis on transcendence leads to a gulf between God and nature and a neglect of divine immanence. Neo-orthodoxy does not do justice to the theme of continuing creation. Moreover, the absolute dichotomy between humanity and nonhuman nature is dubious today, along with the body/soul dualism often used to support such a dichotomy.

We have seen that the Independence thesis is also supported by proponents of *linguistic analysis*, who say that our various self-contained language systems each have their own distinctive rules and functions. Religious language expresses a way of life through the rituals, stories, and practices of a religious community. Creation stories, in particular, provide a cosmic framework of meaning and practical guidance for living. Science, on the other hand, asks strictly delimited questions in the interest of prediction and control. Stephen Toulmin suggests that extrapolating evolution to support either atheism or theism is an illegitimate mixing of languages.[22] The linguists accept an instrumentalist account of both science and religion; there can be no conflict because neither makes truth claims. As a critical realist, I hold that both fields make statements about reality, though these statements are selective and always revisable. We should seek a coherent interpretation of all experience, not a collection of unrelated languages. I will suggest that the evolutionary view of nature has far-reaching implications for our understanding of human nature and God's relation to nature.

2. Primary and Secondary Causality

Since the days of Thomas Aquinas, many Catholic authors have held that God as primary cause works through the secondary causes that science investigates. Because the two kinds of cause operate at totally different levels, scientific analysis can be carried out on its own terms with no reference to theology, as advocates of Independence maintain. The scientific account is complete on its own level, with no gaps in which God would have to intervene, while the theologian can say that God sustains and makes use of the whole natural sequence. Primary causality represents a different order of explanation in answer to questions very different from those asked by the scientist concerning relationships within the natural world.

The Jesuit scientist William Stoeger holds that *God acts through the laws of nature*, using them as instruments for achieving intended goals. "If we put this in an evolutionary context . . . we can conceive of God's continuing creative action as being realized through the natural unfolding of nature's potentialities and the continuing emergence of novelty, of self-organization, of life, of mind and spirit."[23] God's purposes are built into the potentialities of nature, but God also continues to sustain the whole system and holds it in being. Without God it would cease to exist.

Three convictions underlie Stoeger's position. First, we must respect *the integrity of the created order* and *the integrity of science*. There are no gaps in the natural order, and no special intervention was required for the appearance of life or consciousness. Quantum indeterminacies are not gaps that God needs to fill; the laws of nature are patterns of regularity, but they do not imply strict determinism. Second, *God's transcendence and radical otherness* should be acknowledged. God is not a cause like other causes, but is an eternal being in a very different realm who remains a mystery to us. Third, *the creation of persons* was central in God's purposes, and God can use distinctive means of revelation to persons, especially through the person of Christ and our experience of forgiveness and reconciliation. Stoeger leaves open the possibility that God directly communicates information of particular significance in the life of persons.

A similar position is taken by the physicist Howard Van Till, who stands in the Reformed tradition. In replying to Phillip Johnson he insists that *evolutionary science* must be carefully distinguished from both *philosophical naturalism* and *Christian theism*. He says that Johnson is mistaken in claiming that evolutionary theory implies atheism. Van Till cites the early church fathers Basil and Augustine, who held that God did not create everything directly in its present form but gave the world the power to actualize in the course of time an array of specific life-forms. There were no gaps or deficiencies that had to be remedied by subsequent intervention. Of course, these early authors did not have a fully evolutionary scheme, but they did assume that God had empowered created beings themselves with the capacity to realize diverse forms. Augustine held that beings in the world were endowed with "seed principles" from which they could develop the forms conceived for them in the mind of God.[24]

Van Till extends these classical ideas in defense of *the integrity of the created order*. He holds that God created a world pregnant with potentialities that will be actualized in due time, without further divine intervention. Nature was created as a developmental economy without gaps or deficiencies that had to be remedied later. Moreover, God should not be invoked in scientific accounts, which can deal only with relations between members of the created world. Methodological naturalism recognizes the limitations of science, while philosophical naturalism does not. Van Till holds that God provided the rich array of possibilities for viable structures and the pathways among genetic configurations through which they developed, but the scientist can investigate only the structures and pathways themselves. A purposeful pattern can be discerned only in a context wider than science. But the absence of gaps does not imply that the world is closed to divine action, as deism assumed. Special revelatory or redemptive acts of God are possible in this framework.

This is an appealing position because it shows great respect for science while maintaining many of the doctrines of *classical theism*. It tries to avoid deism by asserting that God has a continuing role in sustaining the natural order. But I suggest that it does not fully represent the biblical idea that God has a more active and responsive role in nature and history. Moreover, if all events are predestined in the divine plan, then chance and human freedom are ultimately illusory, though they seem real to us from our limited perspective. Alternatively, if the future is open even in God's sight, we would have to say that not all secondary causes are instruments of God's will. In that case, creation expresses a greater self-limitation in God's power as well as in God's knowledge than classical theism acknowledges (see Chapter 6).

DIALOGUE

Dialogue goes beyond the Independence thesis in exploring conceptual parallels between evolutionary theory and theological doctrines. We will examine three scientific concepts for which theological analogues have been proposed: first, complexity and self-organization; second, the communication of information; and third, top-down causality between levels. These concepts must be considered within science itself before the parallels in theology can be considered.

I. Complexity and Self-Organization

One of the most puzzling scientific questions is how the complex organic molecules necessary for life could have arisen before the presence of reproductive processes through which Darwinian evolution occurs by variation and natural selection. Theories of complexity deal with the inanimate world, but they suggest that higher levels of order can emerge there through *self-organizing systems*. These theories carry further the themes of holism and indeterminacy that we encountered in quantum physics.

Most physical and chemical systems will return to the most probable, disordered state of equilibrium if disturbed from it. But sometimes, if a system is unstable and far from equilibrium, a new level of collective order will appear and achieve a stable form. Ilya Prigogine won a Nobel Prize for his work on dynamic systems far from equilibrium. One of his examples is the sudden appearance of a vortex in the disordered turbulence of a flowing river. Another is the appearance of a complex pattern of convection cells in the circulation of a fluid heated from below. In such cases a small fluctuation is amplified and leads to a new and more complex order that is resistant to further fluctuations and maintains itself with a throughput of energy from the environment. Sometimes there is a "bifurcation of paths" (for example, the convection cells in a heated fluid will all circulate either clockwise or counterclockwise). The choice of paths seems to be the result of very small chance fluctuations.[25]

Prigogine has analyzed many inanimate self-organizing systems in which disorder at one level leads to *order at a higher level*, with new laws governing the behavior of structures showing new types of complexity. Randomness at one level leads to dynamic patterns at another level. In some cases the new order can be predicted by considering the average or statistical behavior of the myriad components. But in other cases, Prigogine shows, there are many possible outcomes, and no unique prediction can be made. Multiple divergent solutions arise from these nonlinear instabilities. The formation of such self-organizing, self-perpetuating systems at the molecular level was perhaps the first step in the emergence of life. As in quantum theory, there seems to be an interplay of law and chance; here, too, we must look at larger wholes and higher levels of organization, and not just at the component parts. Once again, determinism and reductionism are called into question.

Similar conclusions are suggested by *self-organization* in other types of complex systems. Stuart Kauffman finds common patterns in the integrated behavior of systems that appear very different, such as molecules, cells, neural networks, ecosystems, and technological and economic systems. In each case feedback mechanisms and nonlinear interactions make cooperative activity possible in larger wholes. The systems show similar *emergent systemic properties* not present in their components. Kauffman gives particular attention to the behavior of networks. For example, an array of 100,000 lightbulbs, each of which goes on or off as a function of input from its four neighbors, will cycle through only 357 states from among the astronomical number of possible states. Networks are also present among genes, cells, enzymes, and neurons.[26] Many of Kauffman's ideas are speculative and exploratory, but they lead to a new way of looking at prebiotic molecular evolution and the origins of life. He finds that order emerges spontaneously in complex systems, especially on the border between order and chaos. Too much order makes change impossible; too much chaos makes continuity impossible. Complexity at one level leads to simplicity at another level. Disorder is often the precondition for the appearance of a new form of order.

2. The Concept of Information

Information has been an important term in many fields of science. With the advent of computers, instructions can be accurately encoded in a binary representation (0/1 or off/on) and quantified as "bits" of information. The computer responds to the instructions in the program that specify the connections in its electrical circuits. It manipulates the electrical representations of the symbols fed into it ("information processing") and then activates some form of output.[27]

Information is *an ordered pattern* that is one among many possible sequences or states of a system (of alphabetical letters, auditory sounds, binary digits, DNA bases, or any other combinable elements). Information is *communicated* when another system (reader, listener, computer, living cell, etc.) responds selectively—that is, when information is coded, transmitted, and decoded. The meaning of the message is dependent on a wider *context of interpretation*. It must be viewed dynamically and relationally rather than in purely static terms as if the message were contained in the pattern itself.

In organisms, information flows in two directions: both from and to the genes. In *the expression of the DNA* in the growing embryo, the linear message of the DNA molecule produces a linear protein chain, but because the chain has characteristic bonding angles and folds, the result is a distinctive three-dimensional protein structure. Message leads to structure, and structure leads to function. A very complex set of genetic regulatory programs with activators and repressors switches the activity of gene systems on and off, so that the right kind of cell is produced at the right place and at the right time in the growing embryo and in the continuing functioning of the organism. Through chemical feedback signals and time delays, information is fed back to control the genetic sequences. A gene contains an immense number of possible developmental scenarios, of which only a few are realized. In *The Ontogeny of Information*, Susan Oyama argues that the meaning and informational significance of genetic instructions depend on what cells and tissues are already present and on the actual functioning of the developmental system. In place of a one-way flow of information we must imagine interactive construction in a particular context.[28]

DNA constitutes a *developmental* and *functioning* program only in conjunction with molecules in the cytoplasm, which provide a milieu and support structure. The genetic program has been preserved from the past and is effective in the present because of the behavior of larger units—including, finally, the whole interdependent ecosystem, with its cycles and interactions of energy, materials, and information. Each unit achieves stability by being nested in a larger whole to whose stability and dynamism it contributes. As Jeffrey Wicken puts it, "Nature produces itself hierarchically—one level establishing the ground of its own stability by using mechanisms made available by lower levels, and finding functional contexts at higher levels."[29]

Information about the environment has been transmitted to the genes through natural selection. The genes are a record of historically acquired information on what has proved viable and how the organism can make its way through the world, including encoded instinctive behavioral patterns. For example, a bird or animal uses specific visual or auditory cues to recognize and respond to a dangerous predator that it has not previously encountered. Individuals in some species are programmed to communicate warning signals to alert other members of the species. The memory capacity of DNA allows long-term trial-and-error testing in a series of information-gathering experiments reaching up to larger units: organisms, populations, and ecosystems. DNA is part of a larger cybernetic or feedback system for gaining, storing, retrieving, and using information.

Perception is the selective transmission of information about the environment. Even elementary sense organs can detect features of the environment relevant to the organism's life. Perception is an active process in which patterns important to survival are picked out and organized. A one-celled paramecium has a crude perceptual system and a rudimentary form of memory. If it finds no food at one location, it will not persist there but will use its coordinated oarlike hairs to move to another location. Short-term memory requires a new way of storing and recalling information, different from information storage in the genes. Further up the scale, *sentience* seems to involve an internal dimension, an elementary awareness, feeling, and capacity for pain and pleasure. These capacities were presumably selected for their survival value. Pain serves as an alarm system for action to avoid harm. By the time a central nervous system appeared, there was a coordinating network and a new level of integration of experience, which developed eventually into consciousness and finally self-consciousness.

Higher primates are capable of *symbolic communication* of information, and human beings can use words to express abstract concepts. Human information can be transmitted between generations not only by genes and by parental example, but also in speech, literature, art, music, and other cultural forms. The storage and communication of information is thus an important feature of biological processes at all levels, and it must always be understood dynamically and relationally rather than in static and formal terms. Information is communicated only when it is responded to in some way, so it is always context-dependent.

Some theologians have expressed interest in the concept of information. In Chapter 2 we saw that the biblical idea of divine Word (*logos*) involves the communication of meaning and structure. As in the biological world, the meaning of the message has to be decoded and interpreted in a wider context. John Polkinghorne and others have argued that God influences evolutionary history by the communication of information in ways that do not violate the conservation of energy or other scientific laws. We will consider this proposal in Chapter 6 (see "God as Communicator of Information").

3. A Hierarchy of Levels

A feature of biological systems that has been of interest to theologians is top-down causality between levels. Living organisms exhibit a *many-leveled hierarchy* of systems and subsystems. A level identifies a unit that is relatively integrated, stable, and self-regulating, even though it interacts with other units at the same level and at higher and lower levels. One such hierarchy of levels is identified structurally: quark, nucleus, atom, molecule, macromolecule, organelle, cell, organ, organism, and ecosystem. Other hierarchies are identified functionally: the reproductive hierarchy (gene, genome, organism, and population) and the neural hierarchy (molecule, synapse, neuron, neural network, brain, and body). Human beings also participate in all the social and cultural interactions studied by the social sciences and humanities. A discipline or field of inquiry focuses attention on a particular level and its relation to adjacent levels.

We can distinguish three kinds of *reduction between levels*.

1. *Methodological Reduction.* This is a research strategy: the study of lower levels in order to better understand relationships at higher levels. Analysis of molecular interactions has been a spectacularly successful strategy in biology, but it is not incompatible with multilevel analysis and the study of larger systems.

2. *Epistemological Reduction.* This is a relation between theories: the claim that laws and theories at one level of analysis can

be derived from laws and theories at lower levels. I have argued that biological concepts are distinctive and cannot be defined in physical and chemical terms. Distinctive kinds of explanation are valid at differing levels. But interlevel theories may connect adjacent levels, even if such theories are not derivable from the theories applicable to either level alone. The concepts used at any level may change over time as interlevel theories are developed. A series of overlapping theories and models unifies the sciences without implying that one level is more fundamental or real than another.

3. *Ontological Reduction.* This is a claim about the kinds of reality and the kinds of causality that exist in the world. It is sometimes asserted that an organism is "nothing but organized molecules," or that "only physical forces are causally effective." I have defended *ontological pluralism,* a multileveled view of reality in which differing (epistemological) levels of analysis are taken to refer to differing (ontological) levels of events and processes in the world, as claimed by critical realism. In evolutionary history, novel forms of order emerged that not only could not have been predicted from laws and theories governing previously existing forms, but also gave rise to genuinely new kinds of behavior and activity in nature. We can acknowledge the distinctive characteristics of living organisms without assuming that life is a separate substance or a "vital force" added to matter, as the vitalists once postulated.[30]

The presence of a hierarchy of levels has been important in *evolutionary history*. Each level represents a relatively stable unit that preserves its own identity even as it interacts with other, similar units. Modular construction facilitates the assembly of various higher-level configurations without starting each time from the most elementary components. Past advances in complexity can be conserved and used in new ways. Such a system has metastable states, like those of a ball on a hill with terraces on which it can rest if it is disturbed without always returning to the bottom.[31]

Holism is another feature of the rejection of ontological reductionism. Attention is directed to a particular whole, even though it is in turn a part of a larger whole. The whole/part distinction is structural and spatial (e.g., a *larger* whole). *Top-down causality* is a very similar concept, but it draws attention to a hierarchy of many levels characterized by qualitative differences in organization and activity (e.g., a *higher* level). Levels are defined not by size but by functional and dynamic relationships. Patterns in time are emphasized, though of course they are inseparable from patterns in space.

Bottom-up causation occurs when many subsystems influence a system. *Top-down causation* is the influence of a system on many subsystems. Higher-level events impose boundary conditions on chemical and physical processes at lower levels without violating lower-level laws. The state of the upper-level system is specified without reference to lower-level variables. Network properties may be realized through a great variety of particular connections. Correlation of behaviors at one level does not require detailed knowledge of all its components. The rules of chess limit the possible moves but leave open an immense number of possibilities that are consistent with but not determined by those rules. So, too, the laws of chemistry limit the combinations of atoms that are found in DNA but do not determine them. The meaning of the message conveyed by DNA is not given by the laws of chemistry.

Michael Polanyi points out that a machine's design imposes *boundary conditions* on chemical and physical processes. The design does not violate the laws of physics and chemistry, but it harnesses them for organized functions. Polanyi suggests that the morphology and structure of an organism similarly provide boundary conditions that are not required by biochemical laws but are compatible with such laws.[32] Donald Campbell gives a careful analysis of the *top-down causation* through which processes at lower levels are constrained by relationships at higher levels. For example, the huge jaws of the soldier termite are the developmental product of its DNA, but the DNA is itself the product of the selection of the whole organism in its dependence on the termite colony. (The jaws are so large, in fact, that it cannot feed itself and has to be fed by worker termites.) In the world of organisms a complex interaction takes place among many levels.[33] Niles Eldredge and Stanley Salthe speak of an "upward influence" when many lower-level subsystems work together as necessary conditions of a larger whole and a "downward influence" when many subsystems are constrained by the boundary conditions set by higher-level activities.[34] Top-down causality will be important in our discussion of human nature and the mind/body problem in Chapter 5.

The idea of top-down causality has also been extended by theologians who suggest that *God acts as a top-down cause* from a higher level without violating the laws describing events at lower levels. God would be the ultimate boundary condition, setting the constraints within which events in the world occur. We will examine this proposal and compare it with some alternatives in Chapter 6.

INTEGRATION

The last category, Integration, has three versions, as indicated in Chapter 1. Natural theology here takes the form of claims of evolutionary design. A theology of nature is represented by concepts of God's continuing creation through evolution. The systematic synthesis of process philosophy draws extensively from evolutionary ideas.

1. Evolutionary Design

Is evolution a *directional* process? When viewed locally and over short periods, it seems to be characterized by many directions of change rather than progress in one direction. Short-term opportunism is shown when temporarily unoccupied ecological niches are filled even though they may turn out to be blind alleys when conditions change. Gould gives examples in which a structure that originally filled one function was adapted for another function in a makeshift way. For instance, the panda's thumb developed from a bone and muscles in the wrist of its predecessors, a far from perfect design.[35] In some cases we see retrogression, as when a formerly independent organism becomes a parasite. And, of course, by far the majority of species have ended in extinction. The pattern of evolution looks less like a uniformly growing tree than like a sprawling bush whose tangled branches grow in many directions and often die off. Nevertheless, evolutionary history shows an overall trend toward greater complexity, responsiveness, and awareness. The capacity of organisms to gather, store, and process information has steadily increased. Who can doubt that a human being represents an astonishing advance over an amoeba or a worm? Could all this be the product of chance?

Fred Hoyle and Chandra Wickramasinghe argue that the origination of any particular protein chain by chance is *inconceivably improbable*. Suppose you were assembling amino acids to form a hundred-link chain. There are twenty different amino acids from which to choose every time you add a link to the chain. If you assembled chains at random a billion times a second, it would take many times the history of the universe to run through all the possible combinations. These authors suggest that to hope for a particular set of interacting proteins to be produced by chance would be like hoping to make a complete airplane by stirring up a heap of metal parts in a junkyard.[36]

The argument is dubious, however, because there are specific attractive forces between amino acids, and the various combinations are not equally probable or equally stable. As larger structures are formed, stable combinations at various levels will stay together. Complexity comes into being by hierarchical stages, not in one gigantic lottery. Once reproductive processes take place, natural selection is an antichance agency, preserving highly improbable combinations through successive generations. Evolution thus shows a subtle interplay of chance and law.

Of course, the theist may say that *God controls the biological events* that appear to us to be random (an assertion similar to the claim that God controls indeterminate quantum events, discussed in the previous chapter). Chance is pervasive in evolution, including mutations and genetic recombinations. The comet believed to be responsible for the extinction of the dinosaurs could not have been predicted from evolutionary history. Evolutionary history is irreversible and unrepeatable. However, unlike the quantum case, many of these evolutionary events may have occurred because of the accidental intersection of independent causal chains that were themselves lawful and determinate. Moreover, most mutations are harmful or lethal. Is God responsible for them too? There seem to be too many blind alleys and extinct species and too much suffering and waste to attribute every event to God's specific action.

Traditionally, *design* was equated with a detailed preexisting blueprint in the mind of God. The church fathers and subsequent theologians were influenced by the Platonic view of an eternal order of ideas behind the material world. God was said to have a foreordained plan that was carried out in creation. In this framework, chance is the antithesis of design. But evolution suggests another understanding of design—an understanding that postulates a general direction but no detailed plan. A long-range strategy could be combined with short-range opportunism arising from feedback and adjustment. In this view there is increasing order and information but no predictable final state.[37]

D. J. Bartholomew points out that human beings can *use chance to further their purposes*. We toss a coin in the interest of fairness, and we seek random samples in making representative surveys. Many games combine skill and chance; by shuffling cards we generate variety, surprise, and excitement. In evolution, he says, variety is a source of flexibility and adaptability. Varied populations can respond to changing circumstances better than those that are more uniform, and genetic variation is essential for evolutionary change. Chance and law are complementary rather than conflicting features of nature. Random events at one level may lead to statistical regularities at a higher level of aggregation. Redundancy and thresholds may limit the effects of random events on integrated systems. On this reading, chance would be part of the design, and not incompatible with it.[38]

Today we can think of God as the designer of a self-organizing system. Nature is a many-leveled creative process of law, chance, and emergence. We noted earlier the work of Prigogine and Kauffman on self-organization and the appearance of new levels of order in complex systems. We can see *built-in design* in the constraints that limit the possibilities for stable molecular structures and viable developmental patterns in evolutionary history. A world of hierarchical levels seems to have an inherent tendency to move toward emergent complexity, life, and consciousness. A patient God could have endowed matter with diverse potentialities and let it create more complex forms on its own. In this interpretation God respects the integrity of the world and lets it be itself.

These ideas are similar to those of Stoeger and Van Till concerning primary and secondary causality, but they allow a greater role for chance and for the modification of traditional religious doctrines in the light of science. An attractive feature of this option is that it provides at least partial answers to the problems of death and suffering that were such a challenge to the classical argument from design. Competition and death are intrinsic to an evolutionary process. Pain is an inescapable concomitant of greater sensitivity and awareness, and it provides a valuable warning against external dangers.

In discussing the Anthropic Principle in astronomy, I maintained that *design arguments* taken alone (in the tradition of natural theology) are not conclusive but that they can play a supportive role as part of a theology of nature. Design is what one would expect with an intelligent and purposeful God—though I will suggest that the presence of chance, evil, and human freedom should lead us to modify classical ideas of omnipotence. My main objection to such design arguments is that they leave us with the distant and inactive God of deism, a far cry from the active God of the Bible who continues to be intimately involved with the world and human life.

2. God and Continuing Creation

A *theology of nature* is based primarily on religious experience and the life of a historic religious community. Theological doctrines start as human interpretations of individual and communal experience and are therefore subject to revision. Any understanding of God's ongoing relation to nature reflects a particular view of nature. Articulation of the *continuing creation* motif today must take into account the new view of nature as a dynamic, interdependent, evolutionary process.

Arthur Peacocke has written extensively about *models of God in an evolutionary world*. Among classical models he finds Spirit and Word most suitable for expressing immanent divine creativity. God is the communicator, conveying meaning through the patterns of nature as well as through the person of Christ. Peacocke also uses many striking new images. Noting the unpredictability of evolutionary history, he says God is like the choreographer of an ongoing dance or the composer of a still-unfinished symphony, experimenting, improvising, and expanding on a theme and variations. Peacocke uses other analogies that assign a positive role to chance. For example, chance is God's radar beam sweeping through the diverse potentialities that are invisibly present in each configuration in the world. Chance is a way of exploring the range of potential forms of matter.

Peacocke holds that God has endowed the stuff of the world with creative potentialities that are successively disclosed. The actualization of these possibilities can occur only when suitable conditions are present. Events unfold not according to a predetermined plan but with unpredictable novelty. God is experimenting and improvising in an open-ended process of continuing creation. Peacocke rejects the idea of omnipotence and speaks of the self-limitation of a God who suffers with the world.[39]

Peacocke writes that "the natural causal creative nexus of events *is* itself God's creative action." He holds that processes of nature are *inherently* creative. This might be interpreted as a version of the idea that God designed a system of law and chance through which higher forms of life would slowly come into being, which would be a sophisticated form of deism. But Peacocke also says that God is "at work continuously creating in and through the stuff of the world he had endowed with those very potentialities."[40] The images of an improvising choreographer or composer imply an active, continuing relationship with the world, and Peacocke specifically defends the idea of continuing creation.

In other writings Peacocke makes use of the idea of *top-down causality* within the hierarchy of levels in an organism and suggests that God acts from a still higher level. God would be a top-down cause acting on the world as a constraint or boundary condition without violating lawful relationships at lower levels. Peacocke also extends the idea of *whole/part relations* by considering God as the all-encompassing whole of which natural organisms are parts. In addition, he uses the mind/body relation as an analogy for God's relation to the world. The world might be thought of as God's body, he suggests, and God as the world's mind. Cosmic history can be viewed as the action of an agent expressing intentions. We will consider these proposals further in "God as Top-Down Cause" (see Chapter 6).

3. Process Philosophy

Some authors have used the concepts of process philosophy as a way of integrating scientific and religious ideas. Whitehead himself was influenced by *quantum physics* in his portrayal of reality as a series of momentary events and interpenetrating fields rather than isolated enduring bits of matter. The temporality and holism seen in the quantum world were taken to be characteristic of all entities. Process thought rejects determinism, allows for alternative potentialities, and accepts the presence of chance as well as lawful relationships among events. It shares with *evolutionary theory* the conviction that processes of change are more fundamental than enduring substances and that no absolute line separates human from nonhuman life, either historically or in the world today. An organism of any sort is a highly integrated and dynamic pattern of interdependent events. Its parts contribute to and are modified by the unified activity of the whole. Every event occurs in a context that affects it. The process view is ecological in conceiving of the world as a network of interactions in which every entity is constituted by its relationships.

Process thought envisages two aspects of all events as seen from within and from without. The *evolution of interiority*, like the evolution of physical structures, is characterized by both continuity and change. The forms taken by interiority or subjectivity vary widely, starting from rudimentary memory, sentience, responsiveness, and anticipation in simple organisms and going on to consciousness and self-consciousness in more complex ones. Interiority can be construed as a moment of experience, though conscious experience occurs only at high levels of organization.[41]

We have seen that even an *atom* must be considered as a whole and not simply as a sum of its parts. But an atom repeats the same pattern, with essentially no opportunity for novelty except for the indeterminacy of quantum events. Inanimate objects such as stones have no higher level of integration; the indeterminacy of its individual atoms averages out in the statistics of large numbers. A *cell*, by contrast, has considerable integration at a new level. It can act as a unit with at least a rudimentary kind of responsiveness. In a plant there is some coordination among cells but relatively little overall organization.

I suggested earlier that the approach and avoidance reactions of the paramecium can be viewed as elementary forms of *perception* and *response*. An amoeba learns to find sugar, indicating a rudimentary memory and intentionality. Invertebrates have some sentience and capacity for pain and pleasure. Under stress they release endorphins and other pain-suppressing chemicals similar to those in humans. Purposiveness and anticipation are clearly present among lower vertebrates. The development of a nervous system makes possible a higher level of unification of experience. Some mammals exhibit considerable problem-solving and anticipatory abilities and a range of awareness and feelings. I also noted earlier that evolutionary change can be initiated by the activity of organisms in selecting their own environments (the Baldwin effect). Their novel actions may create new evolutionary possibilities. Organisms participate actively in evolutionary history and are not simply passive products of genetic and environmental forces.

New forms of memory, learning, anticipation, and purposiveness appeared in *vertebrates*. The behavior of animals indicates that they can suffer intensely. Consciousness, like sentience, was selected and intensified because it guided behavior that contributed to survival. In *human beings*, the self is the highest level, in which all of the lower levels are integrated. Humans hold conscious aims and consider distant goals. Symbolic language, rational deliberation, creative imagination, and social interaction go beyond anything previously possible in evolutionary history. Humans enjoy a far greater intensity and richness of experience than other life-forms. Process thought attributes some form of interiority (experience, but not consciousness) to integrated entities at all levels but asserts that interiority takes very different forms at differing levels of organizational complexity. This theme is pursued further at the end of the next chapter.

In process thought God is *the source of order* and also *the source of novelty*. God presents new possibilities to the world but leaves alternatives open, eliciting the response of entities in the world. As source of novelty God is present in the interiority of every event as it unfolds, but God never exclusively determines the outcome. This is a God of persuasion rather than coercion.[42] For Christian process theologians, who draw from biblical theology as well as process philosophy, God is not an omnipotent ruler but the leader and inspirer of an interdependent community of beings, as we will see in Chapter 6.

The themes of the present chapter can be summarized as follows.

1. *Conflict*. The Conflict thesis was represented from opposite sides by evolutionary naturalists and by theistic critics of neo-Darwinism. Their views conflict, yet they both say that a person cannot at the same time accept neo-Darwinism and the God of theism.
2. *Independence*. The Independence thesis was exemplified in the separation of domains and functions of science and religion in the writings of some biologists, neo-orthodox theologians, and linguistic analysts and in the sharp distinction other writers have made between primary and secondary causality in evolutionary history.

3. *Dialogue.* The Dialogue thesis involved proposed conceptual parallels between evolutionary theory and theological doctrines, first in relation to complexity and self-organization, then in the communication of information, and finally in the idea of top-down causality between higher and lower levels. All of these concepts suggest possible ways of talking about God's relation to the world.

4. *Integration.* The Integration thesis took three forms: evolutionary arguments for design (natural theology), evolutionary models of God (theology of nature), and the use of evolutionary ideas in process philosophy. The arguments presented start respectively from science (as evidence for design), from theology (in the life of a religious community), and from metaphysics (the search for the most general philosophical categories for interpreting diverse kinds of experience). As in previous chapters, I find myself most sympathetic to the Dialogue position and to Integration as it occurs in a theology of nature and in process philosophy.

Chapter Five

Genetics, Neuroscience, and Human Nature

*I*n Darwin's day, evolution presented a challenge not only to scriptural literalism and the argument from design, but also to the status of humanity. Since then evidence for human descent from prehuman ancestors has accumulated from many scientific disciplines. From molecular biology we know that gorillas and humans today share more than 99 percent of their DNA, though of course the 1 percent that differs is critical. Anthropologists in Africa have found a variety of fossil forms intermediate between gorillas and humans. *Australopithecus afarensis*, an apelike creature, was walking on two legs some four million years ago. In Ethiopia, bones were found from a short female, dubbed Lucy, who walked on two legs, had long arms and a brain size like that of the great apes, and was (as her teeth show) a meat-eater. *Homo erectus*, one million years ago, had a much larger brain, lived in long-term group sites, made more complicated tools, and probably used fire.

Archaic forms of *Homo sapiens* appeared five hundred thousand years ago, and the Neanderthals were in Europe two hundred thousand years ago (though they were probably not in the line of descent to modern humans). The Cro-Magnons made paintings on cave walls and performed burial rituals thirty thousand years ago. The earliest known writing, Sumerian, is six thousand years old. Techniques for melting metallic ores brought the Bronze Age and then, less than three thousand years ago, the Iron Age. Here we have at least the broad outlines of the evolution of both physiology and behavior from nonhuman to human forms and the beginnings of human culture.[1]

Many species of insects and animals live in complex social orders with differentiated roles and patterns of cooperative behavior. In insects these patterns are for the most part genetically determined; in higher animals there is a greater capacity for learning and individuality. Monkeys have elaborate social structures and patterns of dominance and submission. Dolphins form close friendships and engage in playful activity. In such species some of the information relevant to survival is transmitted socially, passed on by parental example rather than by genes. But humans have additional ways of transmitting information from generation to generation, including spoken language, writing, education, and the institutions of society.

Only human beings are fully capable of language, but chimpanzees can be taught limited forms of *symbolic communication*. Chimps lack the vocal organs (especially the larynx) necessary for articulated speech, but they can be taught to communicate in sign language or with geometric symbols on a computer keyboard. They can combine these symbols into simple sentences. From a few examples, chimps can form general concepts, such as *food* or *tool*, and then assign a new object to the correct conceptual category. They can express intentions, make requests, and communicate information to other chimps. These results are impressive, though still far below the level of a two-year-old child. But they suggest that language ability could have evolved gradually.

Higher animals seem to have a rudimentary *self-awareness*. If a chimp sees in a mirror a mark previously placed on its forehead, it will try to remove the mark. But in human beings there is a self-consciousness that seems to be unparalleled. The greater capacity to remember the past, to anticipate the future, and to use abstract symbols liberates us from our immediate time and place. We can imagine possibilities only distantly related to present experience, and we can reflect on goals going far beyond immediate needs. Humans are aware of their finitude and the inevitability of death, and they ask questions about the meaning of their lives. They construct symbolic worlds through language and the arts.[2]

Darwin himself was eager to make the case for the evolutionary origins of human beings and stressed their similarity to nonhuman forms. More recent research has found both *similarities* and *differences*, so evolutionary descent no longer seems in itself to be such a threat to human uniqueness. During evolutionary history, changes in lines of descent were gradual and continuous, but they added up to dramatic differences in ability and behavior. We can defend the distinctiveness of human self-consciousness, language, and culture, for example. Human beings are capable of intellectual and artistic creativity and personal relationships far beyond anything found among other creatures. We are indeed set apart from the rest of nature, but not in the absolute way that classical Christianity assumed.

There are, however, several new scientific disciplines that present strong challenges to traditional religious beliefs concerning human nature. *Sociobiology* (the study of the evolutionary origins of social behavior) offers evolutionary explanations of human behavior, including human morality. *Behavioral genetics* (the study of the role of genes in human behavior today) has been interpreted by some authors as evidence that we are determined by our genes, which would be incompatible with human freedom as it is usually understood. *Neuroscience* (the study of the neural structure and activity of the brain) has led some scientists to maintain that mental events will be exhaustively explained when we understand the interaction of neurons. *Computer science* (the study of information processing in computers, including research on Artificial Intelligence) has led to the claim that the brain is an information-processing system that functions like a computer. Recent research in these diverse fields will be considered within the fourfold typology we have been using throughout.

CONFLICT

Three challenges to religious views of human nature are explored here. First, reductive materialism defends the thesis that all features of human behavior can in principle be accounted for by laws governing the behavior of matter. Second, sociobiology suggests that human morality developed from behavior in our early ancestors that contributed to the survival of their genes. Third, studies in behavioral genetics have been taken to imply that we are controlled by our genes and that human freedom is illusory. Some theological responses to these challenges are also indicated.

I. Reductive Materialism

In *The Astonishing Hypothesis*, Francis Crick, co-discoverer of DNA, has combined the presentation of data from the neurosciences and an explicitly materialist philosophy. He sees only two philosophical alternatives, a supernatural body/soul dualism or a materialistic reductionism. He equates such dualism with religion, of which he is highly critical, unaware that many contemporary theologians have rejected dualism. The volume opens with this statement:

> The Astonishing Hypothesis is that "you," your joys and your sorrows, your memories and your ambitions, your sense of personal identity and free will, are in fact no more than the behavior of a vast assembly of nerve cells and their associated molecules. As Lewis Carroll's Alice might have phrased it: "You're nothing but a pack of neurons."[3]

On the scientific side, Crick is critical of cognitive scientists for relying on computational models and neglecting neural research. His book is mainly devoted to research on visual processing and visual awareness. He proposes that *consciousness* is a product of the correlation of diverse cognitive systems through electrical oscillations of roughly forty cycles per second. He suggests that the activities of various brain regions are coordinated when these oscillations synchronize the local neuron firings. He does not totally dismiss the subjective character of consciousness, but he does not think that it can be studied by science. "What may prove difficult or impossible to establish is the details of the subjective nature of consciousness, since this may depend upon the exact symbolism employed by each conscious organism."[4]

In the previous chapter we considered Daniel Dennett's view of evolution. In other writings he asserts that "consciousness is the last bastion of occult properties and immeasurable subjective states." He holds that qualia (phenomena as experienced) are vague and ineffable. The self is a linguistic fiction generated by the brain to provide coherence retrospectively among its diverse narratives. Dennett holds that the brain forms tentative interpretations of events, multiple draft scenarios of which we are not aware, competing for dominance. The self is the "center of narrative gravity" of the dominant scenarios. It is a useful fiction that we create to provide order in our lives. In other words, the unity and the continuity of consciousness are illusions. Dennett urges us to reject Descartes' concept of an enduring mind, an observer who unifies our diverse perceptions. There are only unconscious processes unified intermittently by a representation of the self that the brain repeatedly recreates from memories of the past and new scenarios in the present.[5]

Dennett describes *the intentional stance* as the strategy of acting as if other people had intentions. He says that the ascription of intentions is predictively useful, but we do not have to assume that intentional states are ever actually present. Dennett claims that he is an *instrumentalist* or *functionalist* who judges concepts only by their usefulness in describing behavior, without asking about their status in reality. But he accepts a metaphysics of materialism when he asserts that neuroscience will be able fully to explain intentional action. He says that he is not a "greedy reductionist" who expects to explain all higher levels directly in terms of the lowest level, but rather a "good reductionist," expecting to explain any level in terms of the next lower one.[6]

Dennett anticipates physical explanations of all mental events. "According to materialists, we can (in principle!) account for every mental phenomenon using the same physical laws and raw materials that suffice to explain radioactivity, photosynthesis, reproduction, nutrition and growth."[7] He asserts: "You are made of robots—or what comes to the same thing, a collection of trillions of macromolecular machines." Both robots and neural networks, in his view, are capable of the behavior that we call intentional. Dennett states that the scientific method can be equated with rationality, and he holds that religious faith is irrational even though it sometimes serves socially useful functions. The philosophy of reductive materialism defended by both Crick and Dennett clearly conflicts with Western religious beliefs about human nature.

2. Sociobiology and Human Morality

Another challenge to traditional religious thought has come from sociobiology. If evolution is the survival of the fittest, how can we explain apparently *altruistic behavior* in which an organism jeopardizes its own survival? Social insects such as ants will sacrifice themselves to protect the colony. Edward O. Wilson and others have shown that such behavior reduces the number of descendants an individual will have, but it enhances the survival of close relatives who carry many of the same genes. If I share half of my genes with my brother or sister, it will perpetuate my genes if I am willing to protect their reproductive futures, even at some risk to my own life. If I help people to whom I am not related, they will in the future be more likely to help me (reciprocal altruism), but this too will contribute indirectly to the survival of my genes.[8]

Wilson is confident that evolutionary biology will account for *all aspects of human life*. Both religion and ethics will be explained and eventually replaced by biological knowledge: "If religion, including the dogmatic secular ideologies, can be systematically analyzed and explained as a product of the brain's evolution, its power as an external source of morality will be gone forever."[9] In the past, he says, morality has been an expression of emotions encoded in the genes. "The only demonstrated function of morality is to keep the genes intact." But now science can "search for the bedrock of ethics—by which I mean the material basis of natural law." "Empirical knowledge of our biological nature will allow us to make optimum choices among the competing criteria of progress."[10]

The evolutionary philosopher Michael Ruse argues that all values are *subjective* and claims that the fact that we think they are *objective* can be explained by sociobiology. Values are actually human constructions that we project on the world, he says, but in order to take them seriously we have to believe that they are objective. Evolutionary selection has favored the cultural myth of objective values. "Darwinian theory shows that in fact morality is a function of (subjective) feelings, but it shows also that we have (and must have) the illusion of objectivity. . . . In a sense, therefore, morality is a collective illusion foisted upon us by our genes."[11] Ruse says that the belief that God is the source of moral rules makes such rules more socially effective and thus serves a useful biological function. It would seem, however, that Ruse's position is self-defeating, for once the secret is out that ethical norms are a collective illusion, we can hardly expect their social effectiveness to continue.

Authors in the new field of *evolutionary psychology* share many of the assumptions of sociobiology. A cover story by Robert Wright appeared in *Time* magazine with the title "Infidelity: It May Be in Our Genes." Wright claims that adultery is natural because behavior favoring a higher number of genetic descendants was selected in Stone Age society. He says that men are particularly prone to promiscuity because they can proliferate their genes with a relatively small investment in the birth and care of children. Women are more likely to seek men with power and status who can be good providers for their children. Wright also claims that there is a genetic basis for criminality, but he holds that belief in free will is a useful fiction because it allows us to assign responsibility and punishment, which are deterrents to criminal behavior.[12]

The philosopher Holmes Rolston has presented an alternative view that conflicts with such interpretations. He gives many examples of the importance of genes in evolutionary history and in human life today, but he maintains that *cultural evolution* differs significantly from biological evolution. First, cultural innovation replaces mutation and genetic recombination as the source of variability. Such innovation is to some extent deliberate and directional; it is certainly not random. New ideas, institutions, and forms of behavior are often creative and imaginative responses to social problems and crises. Next, in the competition between ideas, selection occurs through social experience; the most useful ideas are retained in a trial-and-error process, but many factors enter into social judgments of success. Here selection is less harsh than biological selection, because ideas can be rejected without the death of the individuals who hold them.

Finally, according to Rolston, *transmission of cultural information* occurs through language, tradition, education, and social institutions rather than through genes. Change is more rapid, deliberate, and cumulative than in the case of biological evolution. Major changes can take place within a few generations (or even within a generation). On the other hand, old ideas can surface again and be revived, so they are not permanently lost, as are the genes of extinct species. Because of these differences, cultural beliefs can override or offset the genetic tendencies inherited from our prehuman and Stone Age ancestors.[13]

Rolston gives particular attention to the *evolution of morality*. He holds that terms such as *selfish genes* and *altruism* in lower life-forms are misleading metaphors because at those levels there are no moral agents with the capacity for choice. He claims that the capacity for morality, but not particular moral judgments, is the product of natural selection—just as there is a genetic basis for the capacity for language but not for particular languages, or a genetic basis for the capacity to reason but not for particular rational arguments. He finds implausible the claims that all human altruism is really covert self-interest or the expectation of future reciprocation or social approval. Such explanations simply do not fit the Good Samaritan in the biblical story, or the life of Mother Teresa, or the person who saves the life of a drowning stranger.

Rolston points out that the sociobiologists themselves subscribe to values that cannot be justified by their own theories. It seems implausible that Wilson's deep concern for endangered species, for example, is only an unconscious and indirect way of maximizing his own genetic fitness. Rolston maintains that the capacity for moral judgment is a product of evolutionary history, but he holds that we must turn to philosophy and theology for the grounds of moral judgments. Rolston's view of the evolution of morality is compatible with Christian views of human nature, whereas Wilson's is not.

3. Genetic Determinism and Human Freedom

Behavioral genetics studies the correlation of genes with behavior in the present rather than in evolutionary history. It has sometimes been claimed that our fate is determined by our genes, or that a person cannot be held responsible for a violent act because "his genes made him do it." Studies of identical twins suggest that for many behavioral traits genetic factors account for roughly half of the variation. In one study comparing the brothers of 161 gay men, 52 percent of identical twins (sharing all their genes) were also gay, but only 22 percent of fraternal twins and 11 percent of adopted brothers were also gay.[14] A study of adopted children found that 2.9 percent had criminal records if neither biological nor adoptive parents had criminal records, 6.7 percent if only biological parents had such records, 12.1 percent if only adoptive parents did, and 40 percent if both sets of parents did.[15]

Other studies report that the percentage of Afro-Americans in prison is nine times that of the white population, and some commentators have concluded that *genetic differences* are responsible for criminal behavior. But this interpretation is highly questionable. Most if not all of the racial differences in prison rates can be attributed not to genetic but to social factors, such as higher unemployment among Afro-Americans and discrimination in arrest and conviction rates (which are six times higher than for others for comparable crimes).[16] Controversial genetic studies have often been publicized by the media without the qualifications expressed in the original reports; likewise, later studies failing to confirm the initial reports have seldom been publicized. Studies of alcoholism are particularly problematic because alcoholism seems to have several forms, each of which is affected by many genes as well as by personal history and cultural environment. Genetic and cultural factors cannot be separated in any simple way, because even twins may seek out different environments, subcultures, and experiences that in turn affect their lives. Nature and nurture are always present together, and neither can be considered in isolation.

But even if we allow for the influence of environment as well as genes, we still have not allowed for *human freedom*. Severe constraints are indeed imposed by nature and nurture. Genes establish a range of potentials and predispositions. Parents and social institutions present us with acceptable patterns of behavior. Freedom does not mean that our actions are uncaused or indeterminate, but rather that they are the result of our motives, intentions, and choices and are not externally coerced. Freedom is self-determination at the level of the person. We are not passive stimulus-response mechanisms but selves who can envision novel possibilities and decide deliberately and responsibly among alternative actions. In the case of well-established habits, changes are not easily made, but they can occur if a person seeks a supportive context, as twelve-step programs for alcoholism have shown. We cannot choose the cards we have been dealt, but we can to some extent choose what we do with them.

While we may be only partially controlled by our genes from the past, we know also that the science of genetics will give us new powers over human life in the future. Some religious leaders are opposed to all *genetic intervention* or *genetic engineering*. They claim that to "tamper with DNA," especially in an attempt to alter human nature, would be "playing God" and transgressing our limits. We will lose reverence for life, they say, if we seek God-like powers to redesign it.[17] Manipulation of genes is said to be another form of technocratic arrogance like that portrayed in the myth of Prometheus, the story of the tower of Babel, and the modern novel about Dr. Frankenstein. I suggest, however, that differing views of genetic intervention do not represent a conflict between science and religion as systems of belief, but rather a conflict between differing ethical judgments about the applications of science. Instead of rejecting all forms of genetic intervention, we need to make distinctions among them.

Many authors have maintained that modification of *somatic cells* (body cells that are not passed on to future generations) should be distinguished from modification of *germ cells* (reproductive cells that *are* passed on). The latter modification would have more long-lasting effects but also greater risks of uncertain and perhaps irreversible consequences. Moreover, intervention aimed at *preventing genetic defects* (particularly those resulting in debilitating or fatal diseases, such as Huntington's disease or cystic fibrosis) should be distinguished from intervention aimed at *enhancing particular traits* (such as intelligence, height, or physical strength). The latter is more dubious scientifically (since most traits depend on many interacting genes rather than a single gene) and also more dubious ethically (since it would perpetuate the cultural biases and personal preferences of particular parents or societies). The line between prevention and enhancement is of course not a sharp one, but the motives are likely to be rather different.[18]

A final distinction must be made between intervention in *human* and *nonhuman* life. The cloning of Dolly, the Scottish sheep whose genes were identical with its mother's, was motivated by the hope of producing proteins for the treatment of human diseases. It is a very different matter when someone proposes using what are still high-risk techniques so that people can have themselves cloned. The human clone and the human gene donor would of course differ more than identical twins because they would grow up in differing generations and environments. But human cloning, like gene modification to enhance traits, would treat people as objects to be manipulated or products to be redesigned. Such measures would put a heavy burden of expectation on the newborn child. The search for a "perfect" child might also change our attitudes toward people with genetic disabilities. The Christian tradition holds that God's love and acceptance of each person is unconditional, and that ours should be also.[19] In any case, the discussion of ethical decisions about the uses of our new genetic knowledge is an expression of our human freedom, which implies the rejection of both evolutionary and behavioral genetic determinism.

INDEPENDENCE

There can be no conflict between scientific and religious assertions about human nature if they are independent and unrelated to each other. In the classical body/soul dualism, the soul is said to be immaterial and inherently inaccessible to scientific investigation. Another version of the Independence thesis is found among recent authors who hold that body and soul are terms in two distinct forms of discourse that serve contrasting functions and provide complementary perspectives on human life. Finally, the biblical concepts of sin and redemption seem to have no direct relationship to science, though they are not contradicted by scientific evidence.

1. Body/Soul Dualism

The body/soul dualism found in later Christianity is not found in the Bible itself. In the Hebrew scriptures, the self is a unified activity of thinking, feeling, willing, and acting. H. Wheeler Robinson writes, "The idea of human nature implies a unity, not a dualism. There is no contrast between the body and the soul such as the terms instinctively suggest to us."[20] Oscar Cullmann agrees, noting that "the Jewish and Christian interpretation of creation excludes the whole Greek dualism of body and soul."[21] In particular, the body is not the source of evil or something to be disowned, escaped, or denied—though it may be misused. We find instead an affirmation of the body and a positive acceptance of the material order. Lynn de Silva writes:

> Biblical scholarship has established quite conclusively that there is
> no dichotomous concept of man in the Bible, such as is found in

Greek and Hindu thought. The biblical view of man is holistic, not dualistic. The notion of the soul as an immortal entity which enters the body at birth and leaves it at death is quite foreign to the biblical view of man. The biblical view is that man is a unity; he is a unity of soul, body, flesh, mind, etc., all together constituting the whole man.[22]

According to the *Interpreter's Dictionary of the Bible*, the Hebrew word *nephesh* (usually translated as *soul* or *self*) "never means the immortal soul, but is essentially the life principle, or the self as the subject of appetites and emotion and occasionally of volition." The corresponding word in the New Testament is *psyche*, "which continues the old Greek usage by which it means *life*."[23] When belief in a future life did develop in the New Testament period, it was expressed in terms of the resurrection of the total person by God's act, not the inherent immortality of the soul. Cullmann shows that the future life was seen as a gift from God "in the last days," not an innate human attribute. Paul speaks of the dead as sleeping until the day of judgment, when they will be restored—not as physical bodies or as disembodied souls, but in what he calls "the spiritual body" (1 Cor. 15:44). Such views of the future life may be problematic, but they do testify to the belief that the whole being of persons is the object of God's saving purpose.

However, a dualistic view developed in the early church, largely because of the influence of Greek thought. Plato had held that a preexistent *immortal soul* enters a human body and survives after the death of the body. The Gnostic and Manichaean movements in the late Hellenistic world maintained that matter is evil and that death liberates the soul from its imprisonment in the body. The church fathers rejected Gnosticism but accepted the dualism of soul and body in Neoplatonism and to a lesser extent the moral dualism of good and evil associated with it. Other forces in the declining Greco-Roman culture aided the growth of asceticism, monasticism, rejection of the world, and the search for individual salvation. Some of these negative attitudes toward the body are seen in Augustine's writing, but they represent a departure from the biblical affirmation of the goodness of the material world as God's creation.[24]

In the thirteenth century, Thomas Aquinas accepted the Aristotelian view that *the soul is the form of the body*, which implied a more positive appraisal of the body. He said that the soul was created by God a few weeks after conception, rather than existing before the body. Animals were held to have "sensitive souls," but only humans were said to have "rational souls." Aquinas gave a complex analysis of human nature and moral action that included an important role for emotions ("passions") in carrying out the good, which is known by reason as well as revelation.[25] Medieval theologians expressed a sense of the organic unity of a world designed according to God's purposes. Nevertheless, the concept of an immortal soul established an absolute line between humans and other creatures and encouraged an anthropocentric (human-centered) view of our status in the world, even though the overall cosmic scheme was theocentric (God-centered). With few exceptions, the nonhuman world was portrayed as playing only a supporting role in the medieval and Reformation drama of human redemption.

Descartes' dualism of *mind* and *matter* departed even further from the biblical view. The concept of soul had at least allowed a role for the emotions, as the biblical view had done. But mind in the Cartesian understanding was nonspatial, nonmaterial "thinking substance," characterized by reason rather than emotion. Matter, on the other hand, was said to be spatial and controlled by physical forces alone. It was difficult to imagine how two such dissimilar substances could possibly interact. Descartes claimed that animals lack rationality and are machines without intelligence, feelings, or awareness.

Many theologians have continued to defend a dualism of *body* and *soul*. The official Catholic position is that the human body evolved from the body of primates and proto-human hominids, but the human soul was introduced into a body ready to receive it at a particular point in evolutionary history. In a statement in 1996, John Paul II said that evolution is "more than a hypothesis" since it has been supported by many independent lines of research; he also reaffirmed that throughout human history each soul has been "immediately created by God."[26] Other commentators insist that the soul is immaterial and therefore cannot be discovered by the scientific investigation of either ancient fossils or the brains of present-day humans. They maintain that theological statements about the soul are not derived from scientific research and are quite independent of all scientific theories.

A dualism of *brain* and *mind* has been defended by several prominent scientists. Wilder Penfield points out that the patient whose brain is electrically stimulated is aware that it is not he or she who is raising the arm. Penfield postulates a center of decision radically distinct from the neural network, "a switchboard operator as well as a switchboard."[27] In *The Self and Its Brain*, physiologist John Eccles and philosopher Karl Popper hold that the mind selects among brain modules, reads them out, integrates them, and then modifies other brain circuits: "The self-conscious mind is an independent entity that is actively engaged in reading out from the multitude of active centers in the modules of the liaison areas of the dominant cerebral hemisphere."[28] They point out that impulses appear in the supplementary motor area before those in the motor area only when there is a deliberate, voluntary initiation of action. They defend the interaction of consciousness and brain and the causal efficacy of mental phenomena. But most scientists today do not accept either a body/soul or a brain/mind dualism, though these ideas can still be defended on theological or philosophical rather than scientific grounds.

2. Body and Soul: Complementary Perspectives

Some authors hold that body and soul are not two distinct entities but terms in *two forms of discourse* about a human being. In the tradition of British linguistic philosophy (see Chapter 1), they ask us to examine the distinctive uses of various types of language as they function in human life. To the theologian Keith Ward, reference to soul is a way of asserting human openness to God and the value and uniqueness of each individual. Persons are agents who are both embodied and spiritual. Their actions can be described both in terms of physiological mechanisms and in terms of moral choices, and the two accounts are not mutually exclusive. According to Ward, accounts of brain processes and accounts of conscious mental decisions are derived from different perspectives on the same reality. From one perspective we are biological organisms; from another we are responsible agents created to know and love God. In a religious framework we can defend the dignity and sacredness of human life and the obligation to treat other people as ends in themselves and not just as means to our ends.[29]

A *two-language approach* is also adopted by several respected psychologists with strong theological interests. Malcolm Jeeves holds that mind and brain are two ways of talking about the same events. He cites Donald MacKay's claim that the first-person agent's story of mental events is complementary to (rather than competitive with) the third-person observer's story of neural events. Jeeves says that science and religion also present complementary perspectives, or ways of perceiving the world. In other passages he suggests that there are different levels of activity in the brain to which differing concepts are applicable. These passages go beyond the thesis of two (epistemological) perspectives and suggest that there are two (ontological) aspects of a common set of events, rather than two distinct sets of events.[30]

Fraser Watts holds that "we need to see science and religion as potentially complementary perspectives on the world." He cites the priest Lewis Hall in Susan Howatch's novels as a person who uses the complementary approach in "telling both psychological and religious stories" about the same set of events in a person's life. Watts suggests that "soul talk" captures what we want to say about the uniqueness of persons and their capacity for union with God and potential immortality. But he also hopes that in the future there may be fruitful interaction between theology and the human sciences. He asks whether science and theology might be able to help each other by guiding the construction of theories or by influencing the choices between theories that are underdetermined by the data.[31] Such a proposal would take us beyond the Independence thesis into a more active dialogue.

3. Sin and Redemption

The biblical assertion that humanity is created "in the image of God" (Gen. 1:27) has sometimes been taken to refer to particular human traits, such as rationality, free will, spirituality, and moral responsibility, that distinguish us from other creatures. An alternative view in the history of Judaism and Christianity has been that the *Imago Dei* refers to the relation of human beings to God and indicates their potential for reflecting God's purposes for the world. Human creativity can be seen as an expression of divine creativity.[32] If *Imago Dei* refers to a relationship to God, we would not expect it to be an object of scientific investigation.

But the biblical tradition has also said that we all fall short of fulfilling our creative potentialities. In the light of evolutionary history, the fall of Adam cannot be taken literally. There was no Garden of

Eden, no original state of innocence, free of death and suffering, from which humanity fell. The fall can be taken as a powerful symbolic expression of *human sinfulness*, where sin is understood as self-centeredness and estrangement from God and other people—and, we might add, from the world of nature. Sin in all its forms is a violation of relationships.[33] Original sin is not an inheritance from Adam, then, but an acknowledgment that we are born into sinful social structures, such as those that perpetuate racism, oppression, and violence. Every group tends to absolutize itself, blind to the rationalization of self-interest. Social injustice as well as individual greed is contrary to God's will.[34]

Perhaps at least some aspects of *sin* can be empirically verified. The tragic character of history is evident in its cycles of violence and counterviolence. The massacre of six million Jews in the Holocaust was surely an unmitigated evil. Psychologists can also study empirically the phenomenon of *guilt* in individual life, in both its healthy and its pathological forms. There is some evidence from psychotherapy that too negative a view of human nature and too low an estimate of oneself can be harmful. Guilt without forgiveness, or self-hatred without self-acceptance, seems to hinder rather than encourage love of others. Some theologians join psychologists in calling for a self-respect that is not self-absorption. Perhaps the goal is to seek self-understanding and realism in recognizing both our creative and our destructive potentialities.

Redemption has sometimes been identified with life after death, but it also has a present dimension that is emphasized by many contemporary authors. Redemption is the restoration of relationships—with God, with other people, and with other creatures. Redemption occurs when brokenness and alienation are replaced by wholeness, healing, and reconciliation. The Christian tradition holds that this redemptive possibility is seen most clearly in the life of Christ and in our response to God's love made known in Christ. The doctrine of the Incarnation affirms Christ's full embodiment, underscoring again the importance of the body; it also affirms his unique relationship to God and his total identification with God's will. *Imago Dei*, sin, redemption, and Incarnation can thus all be understood in relational terms rather than as attributes or states of individuals in themselves.

I do not myself accept either the classical body/soul dualism or the proposal that *body* and *soul* are terms in complementary languages. I will defend an integral view of the person as a psychosomatic unity, which I believe is closer to both the biblical view and the evidence from contemporary science. I will suggest that even the biblical concepts of sin and redemption, which seem far removed from any scientific data, should be reinterpreted today in the context of evolutionary history and the social and behavioral sciences.

DIALOGUE

Body/soul dualism has been criticized by both theologians and scientists, and several alternatives have been proposed that allow constructive interaction rather than conflict or independence. The three topics of Dialogue that we will deal with here are neuroscience and the embodied self, anthropology and the social self, and the comparison of the computer and the brain.

1. Neuroscience and the Embodied Self

Many theologians today have tried to reclaim the biblical view of the self as a unified activity of thinking, feeling, willing, and acting. The Bible recognizes the importance of *emotion* as well as reason. "You shall love the Lord your God with all your heart, and with all your soul, and with all your mind" (Matt. 22:37). According to biblical scholars, these three terms—*heart, soul,* and *mind*—describe differing but overlapping human characteristics and activities rather than distinct faculties or components of the person. "The widely held distinction between mind as seat of thinking and heart as seat of feeling is alien from the meaning these terms carry in the Bible. . . . [T]he heart is the seat of the reason and will as well as of the emotions."[35]

Paul writes: "If I understand all mysteries and all knowledge, and if I have all faith, so as to remove mountains, but have not love, I am nothing" (1 Cor. 13:2). Love is of course not simply a matter of emotion, since it involves intention and action. But clearly it is not primarily a product of reason. Some portions of the Bible, such as the Wisdom literature, express the outlook of the wise person reflecting on human experience. But in most biblical texts we are called to be *responsible agents* rather than simply rational thinkers. Sin is understood as a defect of the will, not of reason. In much of Greek thought, the basic human problem is ignorance, for which the remedy is knowledge, while in biblical thought it is our attitudes and motives that lead us astray.

Many *feminist theologians* today are critical of the mind/body dualism. In Chapter 1 I mentioned their critique of the correlations in our culture between the dichotomies of mind/body, reason/emotion, objectivity/subjectivity, domination/nurturance, and male/female. Male is associated with mind, reason, objectivity, and domination, which are given higher status than body, emotion, subjectivity, and nurturance. Feminists decry the denigration of the body in much of Christian history; they seek a more positive evaluation of embodiment and a more integral view of the person. *Environmentalist theologians* criticize the body/soul dualism because it established an absolute line between human and nonhuman life, contributing to attitudes toward other forms of life that have been ecologically destructive.

The importance of the body in mental events is evident in a variety of fields of scientific research. *Perception* itself is an evolutionary product of bodily action. Humberto Maturna and Francisco Verela show that the perceptual systems of organisms do not provide exact representations of the world; they construct information relevant to the needs and actions of the organism. In a frog's visual system some neurons respond only to small dark spots—undoubtedly an advantage in catching flies. So, too, human neurophysiology evolved in parallel with distinctive human goals and interests.[36] Michael Arbib argues that perception is not passive reception of data. Mental representations ("schemas") of the world provide information relevant to possible actions effected through motor programs. Bodily actions are guided by perceptions, expectations, and goals.[37]

The dependence of mental states on *biochemical processes* is shown in the effects of hormones, "mind-altering drugs," and therapeutic medications. For example, Paul Kramer examines the use of Prozac in the treatment of depression. He defends its value in correcting chemical imbalances (especially in the neurotransmitter serotonin), but he concludes that the most effective therapy combines medication with conversation about traumatic experiences and psychosocial factors in the patient's personal history.[38]

The role of *emotions* in cognition is also widely acknowledged by neuroscientists. Antonio Damasio has studied people who have undergone damage in the prefrontal cortex. In a classic case, Phineas Gage recovered from a severe injury and retained his intellectual abilities but underwent a personality change in which he was unable to make decisions or observe social conventions. Another patient with a pre-frontal brain tumor was totally detached emotionally. When he viewed films depicting violence, he could describe appropriate emo-tional reactions but said he could not feel them, and he was unable to make decisions in daily life. Damasio argues that impairments in rea-son and emotion are not separable. He suggests that both Descartes and modern cognitive scientists have neglected the role of emotion in cognition. Damasio also holds that consciousness and continuity of identity are provided by self-representation and the construction of a narrative that includes personal memories and intentions. He describes the self as a many-leveled unity. "The truly embodied mind does not relinquish its most refined levels of operation, those consti-tuting soul and spirit."[39]

In sum, evidence from neuroscience does not support a body/soul dualism, but it does seem to be consistent with theological affirma-tions of embodiment and the importance of emotions.

2. Anthropology and the Social Self

In the biblical tradition, we are inherently social beings. God's covenant was with a people, not with a succession of individuals. Some of the psalms and writings of the later prophets focus on the individual (for example, Jeremiah speaks of a new covenant written in the heart of each person), but individuals are always seen as *persons-in-community*. Judaism has preserved this emphasis on the commu-nity, whereas Protestant Christianity has tended to be more individu-alistic. In the Bible, we are not self-contained individuals; we are constituted by our relationships. We are who we are as children, husbands and wives, parents, citizens, and members of a covenant people. God is concerned about the character of the life of the com-munity as well as the motives and actions of each individual.[40] The religious community shares a common set of sacred stories and ritu-als. Even the prayer and meditation of individuals take place within a framework of shared historical memories and assumptions.

The theme of *the social self* is prominent among contemporary theologians. H. Richard Niebuhr defends the fundamentally social character of selfhood. "Every aspect of every self's existence is conditioned by membership in the interpersonal group."[41] Niebuhr draws from George Herbert Mead and social psychologists who say that selfhood arises only in dialogue with others in a community of agents. We are not impartial spectators but members of communities of interpreters. The social context is also evident in the idea of *the narrative self*. Alisdair MacIntyre and others maintain that our personal identities are established by the stories we tell, the narratives of which we are each the subject. These stories always involve other people.[42] Advocates of "narrative theology" insist that our personal stories are set in the context of the stories of a community. They hold that religious beliefs are transmitted not primarily through abstract theological doctrines but through the stories that provide the wider framework for our own life stories.[43]

Research in anthropology and psychology provides evidence of *the social character of selfhood*. The stories we tell about ourselves as agents and subjects of experience are part of our self-identity. Children learn mental terms and self-referential language as their parents ascribe intentions, desires, and feelings to them. We have a continuing identity as subjects, but memory is always an active reconstruction rather than simply a retrieval of information. We seek coherence and plausibility in our stories; narratives are revised and related to future goals and plans. The tragedy of Alzheimer's disease is the loss of the long-term memory that is required for self-representation. Oliver Sachs describes the case of "the lost mariner" with a brain lesion and memory loss, for whom art and music aided the reconstruction of a new identity.[44]

Emotions in both animals and humans are products of *evolutionary history*. Anger, for example, was presumably selected because it contributed to survival (by mobilizing physiological systems for self-defense and by signaling to opponents a readiness to act). Emotions are not simply instinctive responses, however, for they often require the cognitive appraisal of a situation and a judgment about its potential danger. In human life, culture enters into the social construction of emotions.[45] Emotional feelings are shaped by cultural meanings learned in infancy and throughout life. When children learn words for emotions and culturally approved behaviors to express them, the emotional experience itself is affected. Historians and social psychologists have described the role of emotions as a means of social control (by guilt and shame, for example, in Puritan New England).[46]

Human beings form *symbolic representations* of the self and the world that are always partial and selective. We seek meaning and order by seeing our lives in a wider context—one that is ultimately cosmic in scope. We identify ourselves with purposes and goals that extend beyond our own lives, temporally and spatially. Religious traditions have provided many of the symbols through which individuals integrate conflicting desires and make sense of their lives in a more inclusive context. In story and ritual people participate in religious communities and share their historical memories and their experiences of personal transformation. These wider symbolic structures of order and meaning are indeed human creations, but I have argued that they are also responses to patterns in the world and in human experience, so they can be critically evaluated and revised.[47] Here the religious traditions and the behavioral sciences join in asserting the social character of selfhood.

3. The Computer and the Brain

Many computer experts hold that the human brain is essentially an *information-processing system* that functions like a computer. This view could have been discussed under the Conflict thesis, but a critique of it will lead us into other views that are more open to dialogue. Research in Artificial Intelligence (AI) has the double goal of designing intelligent machines and understanding human cognitive abilities. The dominant view among AI researchers is *symbolic AI* (also called formalist AI, or in some versions, strong AI), which includes four assertions:[48]

1. *The Formalist Thesis.* Intelligence consists in the manipulation of abstract symbols according to formal rules.
2. *The Turing Test.* A computer is intelligent if in performing tasks (such as solving problems, playing chess, or answering questions) it exhibits behavior that we would call intelligent if it were performed by a human being.
3. *Substrate Neutrality (or Multiple Realizability).* Information can be processed in differing physical systems (transistor-based or neuron-based) with identical results.
4. *The Computational Brain.* The human brain functions like a computer. The brain and the computer are two examples of

devices that generate intelligent behavior by manipulating symbols. In popular parlance, *mind* is to *brain* as *software programs* are to *computer hardware*.

Critics of formalism have said that human language and perception are *context-dependent*. Hubert and Stuart Dreyfus have portrayed the importance of commonsense understanding, background knowledge, and nonlinguistic experience in the interpretation of human language. Linguistic and perceptual understanding, they insist, are active processes, strongly influenced by our expectations, purposes, and interests. They have also emphasized *the role of the body* in human learning. Much of our knowledge is acquired actively through interaction with our physical environment and other people. We learn to ride a bicycle not by studying physics or by acquiring a set of rules, but by practice. We use the skills of "knowing how" rather than the propositions of "knowing that." Such "tacit knowledge" cannot be fully formalized. In a child's development, growth in perception is linked to action and bodily movement. These authors see in the formalist thesis the legacy of a rationalism that goes back to Plato: the assumption that knowledge consists of formal rational relationships that exist independent of the body and the material world. They claim that formalism is a new kind of dualism in which software and hardware, like mind and body, can be analyzed independently.[49]

More recently, a set of assumptions different from those of symbolic AI has been used in the design of computerized robots that are *embodied* and *interactive*. The robots in Rodney Brooks's lab at MIT can move around and interact with the world through perception (visual, auditory, and tactile sensors) and action (motion of flexible joints), and they are situated in particular environments. Such robots learn by doing, not by manipulating abstract symbols. Their mechanical bodies are of course very different from our biological bodies; what they learn from their actions differs from what we learn from ours.[50] Anne Foerst, who has degrees in both theology and computer science, works with a group designing a humanoid robot named Cog. She describes four of its characteristics:[51]

1. *Embodiment.* The group holds that human intelligence cannot be separated from bodily action or reduced to computa-

tional abilities. Cog has a "head" and "hands" that can move and interact with its environment.

2. *Distributed Functions.* Small independent processing units activate local motor controls. Modular units with loose connections between them, rather than large centralized programs, allow greater flexibility in coordination and facilitate the acquisition of new abilities.

3. *Developmental Learning.* Like a newborn child learning eye-hand coordination, Cog learns to grasp objects by practice. Many of its capacities are developmentally acquired rather than preprogrammed.

4. *Social Interaction.* Cog practices the equivalent of "eye contact" and is programmed to take into account some of the effects of its actions on people. These social features are at an elementary stage but are a goal of ongoing research.

Recent work in robotics answers some of the objections raised against the symbolic AI program, but other questions remain in the comparison of artificial and human intelligence. The *process of socialization* in humans occurs over a span of many years. In computers, information processing is very rapid, but interaction with the environment takes considerable time. Robots might be socialized by being fed vast quantities of information, but if the critics of formalism are correct, participation in human culture and patterns of life would require active interaction over a longer period of time. The Dreyfuses maintain that only computer systems nearly identical with the human brain and endowed with human motives, cultural goals, and bodily form could fully model human intelligence. That may be too strong a claim, but it points to the importance of culture as well as body in human understanding and in any attempt to duplicate such understanding in machines.

Most AI researchers claim only to simulate cognitive processes, and they hold that cognition is quite independent of *emotions*. Roger Shank writes: "It would seem that questions such as 'Can a computer feel love?' are not of much consequence. Certainly we do not understand less about human knowledge if the answer is one way or the other. And more importantly, the ability to feel love does not affect its ability to understand."[52] Some authors suggest that we can analyze the function of an emotion in evolutionary history and then try to construct an AI program that fulfills the same function. For example, the main behavioral function of fear is avoidance of danger, which might be programmed directly. Others hold that computers could not experience feelings but could represent the cognitive components of emotions (such as the external causes of anger and its relation to one's beliefs and ensuing actions).[53]

Rosalind Piccard's research is directed toward building computers with the ability to *recognize and express emotions* in order to facilitate communication between computers and humans. For example, a computer instruction program could slow down or offer further explanations when it perceived expressions of frustration or anger in the user's face or heartbeat. A computer voice synthesizer might deliver a message with an intonation conveying an appropriate emotional tone. But Piccard remains agnostic as to whether future computers might actually experience emotions. If they did, their feelings would be very different from ours, which arise in a complex biological organism.[54]

There are still enormous differences between *computers* and *brains*. A brain has trillions of neurons, each connected to as many as ten thousand neighbors; the number of possible ways of interconnecting them is greater than the number of atoms in the universe. Signals between neurons are not digital but are encoded in continuously variable properties such as electrical potentials or neuron firing frequencies. New knowledge from neuroscience will undoubtedly affect future computer design, but we should not underestimate the differences or the difficulties.[55]

INTEGRATION

Whereas exponents of Dialogue may sometimes reformulate traditional religious ideas on particular topics in the light of science, advocates of Integration draw more extensively and systematically from the sciences and are willing to accept more far-reaching theological reconstruction. Three examples are presented here: theological views of the person as both a biological organism and a responsible self; philosophical analyses of mind and brain as two aspects of one process; and the understanding of selfhood in process philosophy.

I. Biological Organism and Responsible Self

The theologian Philip Hefner holds that we are *created co-creators* in an ongoing process of continuing creation. Evolution is God's way of creating free creatures and thereby opening up further creative possibilities. We are at the same time creatures of nature and culture, conditioned by genes and previous history. But as co-creators, we have freedom and the capacity to seek new directions that are novel and yet are within the constraints set by our genetic and social inheritance. Hefner says that nature is "stretched" and "enabled" as it gives rise to the new zone of freedom. "*Homo sapiens* is God's created co-creator whose purpose is the stretching/enabling of the systems of nature so that they can participate in God's purposes in the mode of freedom."[56] God is immanent in the creativity and self-transcendence evidenced in evolutionary history and continuing into the future.

Hefner maintains that we can participate in *God's ongoing creative work:* "We humans created in the image of God are participants and co-creators in the ongoing work of God's creative activity. We are being drawn toward a shared destiny which will ultimately determine what it means to be a true human being."[57] Hefner holds that Christ is the prototype of true humanity and represents a radically new phase in cultural evolution. In Christ we come to know God's will as universal love. The eschatological hope is a confidence in God's purpose to perfect and fulfill the creation. We human beings can be conscious agents in a new level of creation, says Hefner, but we are also in a stage of great precariousness and vulnerability. Technology gives us immense powers over nature, and our decisions will affect all terrestrial life. We have a responsibility not only for our own future but also for the rest of the creatures on our planet.

Hefner suggests also that an evolutionary interpretation can be given to *the concept of sin.* He identifies sin with the conflict between information in the genes and that in culture. Genetically based selfishness is in conflict with cultural sources of cooperation and altruism. Original sin consists in biologically based dispositions from the past that are not adaptive in the modern world.[58] The problem with this analysis is that it tends to make genes the source of evil and culture the source of good. I would argue that social injustice, violence, racism, and militarism are products of culture and social institutions and of individual decision as much as of inherited genes.

Ronald Cole-Turner asks how our view of human nature affects our decisions about *genetic intervention*. Genetic engineering, he says, is not "tampering with God's creation" (as if nature were a static and completed order) but participating in God's continuing creative activity. God works through natural processes and can work through us if we choose to follow God's purposes. Cole-Turner is aware of the temptation to misuse our powers, as expressed in the traditional idea of sin. He says that we must provide safeguards to limit the concentration of power in biotechnology industries, such as that which occurs when naturally occuring genes are discovered and patented. But genetic intervention can be a significant path to the relief of suffering and a fulfillment of God's purposes. Cole-Turner says that Christ's healing ministry supports the wider claim that not everything in nature is in accord with God's intention.[59]

Cole-Turner maintains that we are *called to participate* in God's creative and redemptive work. We can be creation-affirming and creation-transforming, as God is, when we cooperate in the service of God's purposes. Cole-Turner is aware of the genetic constraints on our lives (including our capacities for religious experience and human love), but he defends our freedom to intervene to correct genetic defects, to reduce human suffering, and to use our technological powers to further social justice and human fulfillment. He emphasizes the positive possibilities for genetic engineering, though he recognizes the dangers in germ-line intervention, especially in the enhancement of human traits. Like Hefner, Cole-Turner combines a recognition of the scientific evidence for the influence of genes on human behavior with a reformulated religious understanding of human sinfulness, freedom, and responsibility.

2. Mind and Brain: Two Aspects of One Process

The reductive materialism of Crick and Dennett and the mind/body dualism of Descartes were described earlier, but these are not the only philosophical alternatives. Owen Flanagan maintains that some features of phenomenal first-person accounts can be correlated with data from cognitive psychology and neuroscience. He takes seriously our conscious experience—that is, our awareness of sensations, perceptions, emotions, beliefs, thoughts, and expectations. Flanagan describes neural correlates of visual experience, such as the neurons that respond to edges, shapes, colors, and motions, and discusses the brain activities that are correlated with the emotions of fear and anger. But he says that high-level concepts of the self are not expressible in neural terms. Human actions, for example, must be identified by the intentions that constitute them.[60]

Flanagan argues that *the self is constructed;* it is not given to us as a single entity or a transcendental ego. The newborn gradually builds an integrated self with the help of parents and other people. With maturation and socialization, a distinct identity is shaped, cast largely in narrative form in the stories we tell ourselves. The self changes as a result of active engagement with the environment and other persons. Our self-representations organize our memories of past events and our plans and aspirations for the future. Models of the self do not use concepts applicable to neurons, for they reflect our aims and values and our patterns of action and human relationships. The narrative self has *causal efficacy* as a complex and ever-changing self-representation. It causes people to say and do things and hence has an ontological and not merely a linguistic status. The self is a many-leveled reality that is constructed rather than given; activities at each level have some autonomy and yet are related to each other.

David Chalmers holds that *consciousness is irreducible* but argues that all other biological and psychological states are determined by physical states and are in principle explainable by physical theories. He holds that the cognitive sciences can give detailed functional accounts of memory, learning, and information processing, but they cannot say why these processes are accompanied by conscious experience, which is not defined by its causal roles. Phenomenal subjective experience is known firsthand in sensory perception, pain, emotions, mental images, and conscious thought.

Chalmers rejects materialism and functionalism and defends *a two-aspect theory,* which he also calls property dualism or a form of panpsychism. He proposes that *information states* are the fundamental constituents of reality and that they are always realized both *subjectively* and *physically:* "We might say that the internal aspects of these states are phenomenal and the external aspects are physical. Or as a slogan: Experience is information from the inside; physics is information from the outside."[61] A dog has access to extensive perceptual information, so we can assume it has rich visual experiences. A fly has rather limited perceptual discrimination and also a lower level of experience with fewer phenomenal distinctions. Simple information states would be realized in simple physical structures and simple phenomenal experiences. "It is likely that a very restricted group of subjects of experience would have the psychological structure required to truly qualify as *agents* or *persons*."[62] Such two-aspect views of the person offer an alternative to both mind/body and body/soul dualisms.

3. Process Philosophy

Whitehead and his followers hold that the basic components of reality are not two kinds of enduring substance (mind and matter), or one kind (matter), but *one kind of event with two aspects or phases*. Whereas substances remain the same in different contexts, events are constituted by their relationships and their contexts in space and time. This philosophy portrays the dipolar character of all unified events, but it recognizes that these events can be organized in diverse ways, leading to an organizational pluralism of many levels. All integrated events at any level have an inner reality and an outer reality, but these take very different forms at different levels. Both the interiority and the organizational complexity of psychophysical systems have evolved historically.

Process thinkers agree with dualists that *interaction* takes place between the mind and the cells of the brain, but they reject the dualists' claim that this is an interaction between two totally dissimilar entities. Between the mind and a brain cell there are enormous differences in characteristics, but not the absolute dissimilarity that makes interaction so difficult to imagine in dualism. The process view has much in common with two-language theories or a parallelism that takes mental and neural phenomena to be two aspects of the same events. But unlike many two-aspect theories, it defends interaction, downward causality, and the constraints that higher-level events exert on events at lower levels. At higher levels there are new events and entities, not just new relationships among lower-level events and entities.[63]

We have seen that in looking at diverse types of systems, Whitehead attributes *subjective experience* in progressively more attenuated forms to persons, animals, lower organisms, and cells (and even, in principle, to atoms, though at that level subjective experience is effectively negligible), but not to stones or plants or other unintegrated aggregates. David Griffin proposes that this should be called *panexperientialism* rather than *panpsychism*, since for Whitehead mind and consciousness are found only at higher levels. Consciousness occurs only in higher forms with a central nervous system. Every entity is a subject for itself and becomes an object for others, but only in higher life-forms are data from brain cells integrated in the high-level stream of experience we call *mind*. Consciousness and mind are thus radically new emergents in cosmic history.

The conceptual system of process philosophy expresses several of the themes developed earlier in this chapter.

1. *Embodiment.* Every event is portrayed in process thought as a synthesis of past and present bodily events. There are no events that have a subjective aspect without an objective aspect. We experience the causal efficacy of our own bodies. The senses, such as sight, always have a bodily reference rather than simply transmitting information about the world.

2. *Emotions.* Process thought recognizes the importance of nonsensory experience and the perception of feeling in our own bodies. Consciousness and cognitive thought occur against a background of feeling. Whitehead writes: "The basis of experience is emotional. . . . The basic fact is the rise of an affective tone originating from things whose relevance is given."[64]

3. *A Hierarchy of Levels.* Among process thinkers, Charles Hartshorne has developed most fully the idea of a sequence of levels intermediate between the atom and the self. He dwells on the differences between cells and mere aggregates such as stones. A holistic outlook directs attention to system properties that are not evident in the parts alone. But process philosophy recognizes that various levels may be integrated according to very different principles of organization, so their characteristics may be very different. In a complex organism, downward causation from higher to lower levels can occur because every entity is what it is by virtue of its relationships. Every entity is influenced by its participation

in a larger whole. Emergence arises in the modification of lower-level constituents in a new context.[65]

4. *Conscious and Unconscious Processes.* Whitehead says that consciousness first appeared in animals with a central nervous system as a radically new emergent. In human beings, most mental activity is unconscious. Consciousness occurs only in the last phase of the most complex occasions of experience, as a derivative byproduct of nonconscious experience. It involves the unification of information from the past and from the body with a new element: the contrast of past and future, the entertainment of possibilities, the comparison of alternatives. Self-identity consists in the continuity of processes, most of which are below the threshold of awareness.

5. *The Construction of the Self.* Whitehead was influenced by William James, who held that there is no enduring self, only the stream of experience. Thought goes on without a thinker, or even a succession of thinkers aware of the same past. Continuity of identity, James said, is guaranteed only by the persistence of memory. Whitehead also holds that the self is a momentary construction, but he asserts that it is a unified, complex process. The unity of self is a unity of functioning, not the unity of a Cartesian thinker. This view that selfhood is constructed is consistent with recent neuroscience. Process theologians argue that it is also consistent with a religious understanding of human nature.

In process thought, God's attributes include distinctive forms of embodiment, emotion, consciousness, and social interaction (see Chapter 6). God is present in all time and space but also transcends time and space. God is eternal and unchanging in character and purpose, but temporal in being affected by interaction with the world. The process view of human immortality, like its view of sin, redemption, and the Incarnation, is relational—that is, immortality is a relationship of persons to God and other beings, not a property of individuals in themselves. Some process theologians defend only *objective immortality*—our effect on God and our participation in God's eternal life—in which evil is transmuted and good is saved and woven into the harmony of the larger whole. Others defend *subjective immor-*

tality, in which the human self continues as a center of experience in a radically different setting, amid continuing change rather than changeless eternity, and with the possibility of continued communion with God.[66] John Cobb speculates that we might picture a future life as neither absorption in God nor the survival of separate individuals, but as a new kind of community transcending individuality.[67] This version of immortality is closer to the classical tradition, but it preserves the process theme of relationality, rather than postulating an inherently immortal soul.

In summary of this chapter, I believe that both recent theology and recent science support a view of the person as a multilevel psychosomatic unity who is at the same time a biological organism and a responsible self. The sciences that I considered were evolutionary biology, genetics, neuroscience, anthropology, and computer science. Some of the levels of human life are shared with all matter, some with all living things, and some with animal life, while some seem to be uniquely human. Human morality is a product of culture, and we can choose to resist some of the tendencies inherited from an evolutionary past. Our behavior today is constrained (but not completely determined) by our genes. Embodiment, emotion, and social relationships are constitutive of selfhood as understood in both biblical religion and current science. Both conscious and unconscious processes occur in mental activities; the brain is indeed an information processor, but it has other characteristics not shared by computers. Both reductive materialism and the dualism of mind and matter (or body and soul) are avoided by two-aspect theories and by the dipolar pluralism of process thought. The self can thus be conceived not as a separate entity but as the person in the unified activity of thinking, feeling, and acting.

Chapter Six

God and Nature

$\mathcal{I}$n previous chapters we have looked at some theological implications of particular sciences, from astronomy and physics to biology, genetics, and neuroscience. In this concluding chapter we ask a broader question concerning God's relation to nature: How can God act in a world of lawful natural processes?

We have seen that the Bible includes a variety of theological models. God is represented as a purposeful designer imposing order on chaos, a potter or craftsman making an artifact, and an architect setting out the foundations of a building. God is said to be a life-giving Spirit at work throughout nature and a communicator expressing meaning and rational structure through the divine Word. God is the sovereign ruling over both history and nature. In relation to Israel, God is the liberator delivering the community from bondage and a judge dedicated to righteousness and justice. In relation to individuals, God is the careful shepherd, the forgiving father, and (more rarely) the nurturing mother. God is also the redeemer who brings new wholeness to communities and individuals—and even to nature in the final fulfillment.

In the Middle Ages, biblical and Aristotelian ideas were brought together, especially in the writings of Thomas Aquinas, which were so influential in later Catholic theology. The biblical model of God as king and ruler was elaborated into formal doctrines of divine omnipotence and omniscience. The dominant model was that of the absolute monarch ruling over his kingdom, though other models were also present. A similar view of God was prominent in the Reformation, particularly in Calvin's emphasis on divine sovereignty and predestination.

In the classical doctrine of *divine omnipotence*, God governs and rules the world in providential wisdom. All events are totally subordinate to God's will and power. Foreordination was said to involve not only foreknowledge but also the predetermination of every event. Both medieval Catholicism and Reformation Protestantism held that God intervenes miraculously as a direct cause of some events, in addition to the more usual action of working through secondary natural causes. This was taken to be a strictly one-way relation: God affects the world, but the world does not affect God, who is eternal and unchanging.[1]

The exclusion of all temporality from God's nature in medieval thought seems to have been indebted mainly to Greek thought. Plato had pictured a realm of eternal forms and timeless truths, imperfectly reflected in the world; the perfect was the unchanging. Aquinas argued that God is *impassible*, unaffected by the world. God loves only in the sense of doing good things for us, but without passion or emotion. God's being is wholly self-sufficient; independent of the world, it receives nothing from the world. Since God knows all events in advance and controls every detail, divine knowledge is unchanging; in God there is no element of responsiveness. In the last analysis, the passage of time is unreal to God, for whom all time is spread out simultaneously. All of this seems to contrast with the dynamic God of the Bible, who is intimately involved in Israel's history and responds passionately to its changing situations.

To be sure, other themes qualified the medieval and Reformation image of divine sovereignty. God's control was never portrayed as sheer power, for it was always the power of love. Dante ended *The Divine Comedy* with a vision of God as "the Love that moves the Sun and other stars."[2] Classical theism indeed emphasized transcendence, and God was said to act occasionally by supernatural intervention from outside nature. However, divine immanence was also defended. God was held to be preeminently present in the Incarnation, the sacraments, and the life of the church, but the Holy Spirit was said to animate nature as well as human life.

Six kinds of objection have been raised to the monarchial model of God as omnipotent ruler:

1. *An Evolutionary World.* During the centuries when the monarchial model was dominant, a static view of reality was assumed. The world was accepted as a fixed order whose basic forms were unchanging, given once for all. This tended to reinforce the idea of creation *ex nihilo* in an absolute beginning; the biblical idea of continuing creation was virtually ignored. Each lower form served the higher in a preordained hierarchy: God/man/woman/animal/plant. This fixed order was unified by God's sovereign power and omniscient plan. These assumptions were challenged by the evolutionary view of the world.

2. *Law and Chance in Nature.* With the rise of modern science, the idea of supernatural intervention in nature seemed increasingly dubious. By the eighteenth century, God's wisdom and power were seen primarily in the initial design of the universe, not in its continuing governance. Deism took seriously the lawfulness of nature at the price of relegating God's activity to the distant past. We have seen that more recently the role of chance has called into question both the determinism of predestination and the determinism of natural law.

3. *Human Freedom.* Divine omnipotence and predestination appear incompatible with the existence of genuine alternatives in human choice. No subtleties in distinguishing foreknowledge from foreordination seem to be able to circumvent this basic contradiction. Humanity's total dependence on and submission to an authoritarian God is also in tension with human responsibility and maturity. If all power is on God's side, what powers are assignable to humanity?

4. *Evil and Suffering.* Why would a good and omnipotent God allow evil and suffering? If we accept evolution, suffering and death cannot be viewed as consequences of Adam's fall. If omnipotence is defended, and everything that happens is God's will, then God seems to be responsible for evil. Many recent theologians have suggested that evil and suffering (as well as human freedom) point to God's voluntary self-limitation rather than to the exercise of omnipotence.

5. *Patriarchy*. The characteristics of the monarchial God are those that our culture identifies as "masculine" virtues (power, control, independence, rationality, and impassibility), rather than virtues stereotyped as "feminine" (nurturance, responsiveness, interdependence, and emotional sensitivity). The identification of God with "masculine" qualities seems to reflect the biases of a patriarchal culture, and this model of God has in turn been used to justify male dominance in society.

6. *Religious Intolerance*. The exaltation of God's power encouraged an exclusivist view of revelation. Taken with a hierarchical understanding of the authority of the church, it was used to support absolute claims to religious truth. When coupled with political and military power, it led to religious persecution, crusades, holy wars, and colonial imperialism, all in God's name. Such views are a continuing danger in a world of religious pluralism and weapons of mass destruction.

We will examine some responses to the challenges to classical theism arising from contemporary science, using the same fourfold typology used in previous chapters. Conflict is evident in various forms of naturalism, which rejects all ideas of a personal God. Defenders of Independence maintain that there can be no conflict because theological assertions are on a totally different level from scientific assertions; theology and science are complementary rather than competing languages. Dialogue is exemplified when scientific concepts (such as self-organization or information) are used to reformulate traditional ideas of God's relation to the world. Finally, some authors seek integration by conceiving of God as determiner of indeterminacies or as top-down cause, or by using the philosophical categories of process theology.

CONFLICT

We first consider some criticisms of the concept of God by naturalistic scientists. We will then examine some forms of religious naturalism in which traditional ideas of God are rejected but responses to dimensions of the natural order that can serve similar functions in human life are affirmed.

I. Naturalistic Critiques of Theism

Sigmund Freud held that religious beliefs are a product of *wishful thinking*. They are an evasion of reality and a regression to childhood in the search for security in a hostile world. Children are dependent on their fathers when they are small, but as they grow older they recognize the limitations of their human fathers and seek a new security in the projection of an idealized father-image on a cosmic scale. Freud's convictions about the importance of childhood experience were combined with a naturalistic philosophy that he claimed was based on science. He maintained great confidence in human reason, despite his ideas about the power of unconscious and irrational forces in human life.[3]

The theist can reply that while the search for security and the evasion of responsibility are all too common in religious history, many prophets, saints, and ordinary people have found their religious beliefs to be a source of courage in taking risks and in working for the welfare of others. Some of our wishes are realistic; others are not. If all beliefs could be dismissed as wishful thinking, naturalism would itself be undermined. The theist can argue that if God is personal, a parent's love at its best might be an appropriate analogy for God's love. The scientific basis for Freud's claims have also been questioned; his patients were a limited group in a particular society, many of them suffering from neurotic guilt in a moralistic culture. His naturalism should be seen not as a scientific conclusion but as a philosophical assumption.

Among contemporary authors, the cosmologist Peter Atkins is one of the most vehement in his *dismissal of theism*. He holds that the universe was the product of chance vacuum fluctuations. In the subsequent thermodynamic collapse into disorder, occasional islands of order were thrown up by purposeless physical processes. Religion is "sentimental wishful thinking" and "intellectually dishonest emotion." "The whole history of the church was based on a clever but understandable self-delusion (and in some cases, I suspect, on a straightforward conscious lie)."[4] By contrast, Atkins gives a triumphalist account of science. "Complete knowledge is just within our grasp. Comprehension is moving across the face of the earth like the sun."[5] He writes:

Reductionist science is omnicompetent. Science has never encountered a barrier it has not surmounted or that we can reasonably suppose it has power to surmount and will in due course be equipped to do so. . . . Religion has failed, and its failures should be exposed. Science, with its currently successful pursuit of universal competence through the identification of the minimal, the supreme delight of the intellect, should be acknowledged king.[6]

In Chapter 4 we discussed Richard Dawkins's assertion that the presence of chance shows that this is *a purposeless universe*. Natural selection, he says, is a mechanism that makes highly improbable molecular combinations probable by the accumulation of many small adaptive improvements.[7] "We are machines for propagating DNA, and the propagation of DNA is a self-sustaining process. It is every living object's sole reason for living."[8] Dawkins takes theism to be a rival theory that competes with science in explaining the world. He presents a challenge to the theist: "Either admit that God is a scientific hypothesis and let him submit to the same judgment as any other scientific hypothesis, or admit that his status is no higher than that of fairies and river sprites."[9]

Elsewhere he writes: "I think a case can be made that *faith* is one of the world's great evils, comparable to smallpox virus but harder to eradicate. Faith, being belief that isn't based on evidence, is the principal vice of religion. . . . Don't fall for the argument that religion and science operate on separate dimensions or are concerned with quite separate sorts of questions. Religions have historically always attempted to answer questions properly belonging to science."[10]

In his recent book *Consilience*, E. O. Wilson carries further the Enlightenment vision of *the unity of knowledge*. He holds that advances in science are beginning to make possible the consilience (unification) of the natural sciences with the social sciences and humanities. The pursuit of interdisciplinary connections is indeed a significant task, and Wilson ranges widely and eloquently over many disciplines. But in his version, unity is to be achieved on terms dictated by science:

I have argued that there is intrinsically only one class of explanation. It traverses the scales of space, time, and complexity to unite the disparate facts of the disciplines by consilience, the perception of a

seamless web of cause and effect. . . . The central idea of the consilience world view is that all tangible phenomena, from the birth of
stars to the workings of social institutions, are based on material
processes that are ultimately reducible, however long and tortuous
the sequences, to the laws of physics. . . . No compelling reason has
ever been offered why the same strategy should not work to unite
the natural sciences with the social sciences and humanities. The differences between the two domains is in the magnitude of the problem, not the principles needed for its solution.[11]

One reviewer writes, "Wilson's model of unity is not collaboration
across disciplines but a hostile takeover of the humanities and social
sciences by the natural sciences."[12]

Wilson offers an *evolutionary account* of religious beliefs, similar to
his evolutionary account of ethics summarized in the previous chapter. He describes the hierarchies of dominance and submission that
have contributed to the survival of animal species. Every wolf pack or
group of monkeys has a dominant male, and the submission of other
members is signaled by facial expression, tail position, and deferential
behavior. Wilson suggests that comparable submissive behavior is
expressed in human deference to charismatic leaders and religious
authorities—and ultimately to a supremely dominant figure, typically
a male God. Submission is enacted and reinforced in ritual and
prayer. "Behavioral scientists from another planet would notice immediately the semiotic resemblance between animal submissive behavior on the one hand and human obeisance to religious and civil
authority on the other. They would point out that the most elaborate
rites of obeisance are directed at the gods, the hyper-dominant if
invisible members of the human group."[13]

Wilson holds that religion has served useful functions in the past
and that people need sacred narratives in which their lives can be
viewed as part of a larger purpose. Today these functions can best be
filled by a poetic rendition of the evolutionary epic. "If the sacred
narrative cannot be in the form of a religious cosmology, it will be
taken from the material history of the universe and the human
species. That trend is in no way debasing. The true evolutionary
epic, retold as poetry, is as intrinsically ennobling as any religious
epic. Material reality discovered by science already possesses more
content and grandeur than all religious cosmologies combined."[14]

2. Religious Naturalism

Under the heading of Conflict several of the previous chapters have included, along with antireligious naturalists, people from the conservative end of the theological spectrum who do not accept evolution or other scientific theories. In this chapter I will instead look at defenders of a religious naturalism that is less reductionistic and more sympathetic to religion than that espoused by the authors considered above. These versions of naturalism retain at least some concepts from the Western religious heritage but radically reformulate them to accommodate contemporary science. Perhaps these authors might wish to be included in our fourth category, Integration, because they do integrate science with their minimalist understanding of religion. But in my judgment they have rejected so many traditional religious beliefs that they should be considered examples of Conflict—despite their thoughtful efforts at reinterpretation and their portrayal of some continuity as well as more evident discontinuity between their views and classical Christianity.

Ralph Burhoe, founder and for many years editor of *Zygon: Journal of Science and Religion*, holds (in contrast to Wilson) that altruism toward individuals who are not close relatives cannot be explained by genetic selection. He contends that religion has been the major force in fostering altruism extending beyond genetic kin. The set of values transmitted by religious myths and rituals binds a society together. Religion has often fostered loyalty to one's group and hostility to other groups that threatened it, which furthered group survival. But as tribal religions gave way to more universal ones, the circle of loyalty expanded. For many centuries each of the major world religions represented "a well-winnowed wisdom" expressed in terms of the best understandings of the world available in its time. But to be credible today, Burhoe holds, religious beliefs must be firmly based on scientific evidence. This scientific grounding will encourage globally shared values, which are needed for survival in the nuclear age.[15]

Burhoe advocates *evolutionary naturalism* as the religious philosophy best suited to a scientific culture. For him, nature has many of the characteristics of the traditional God—especially omnipotence—and it should be the object of our worship and obedience. We are totally dependent on the evolutionary process for our existence, our sustenance, and our destiny. Nature is the sovereign power on which we are dependent; it is our creator and judge. "It makes little difference whether we name it God or natural selection," says Burhoe.[16] We must adapt to the requirements of "the all-determining reality system. . . . Man's salvation comes in recognizing this fact and adapting to it or bowing down before the majesty and glory of the magnificent program of evolving life in which we live and move and have our being."[17]

By *equating God with nature*, Burhoe seems to have abandoned theism. In his late writings he does sometimes express a sense of mystery that might suggest that science has its limits, but he also asserts that nature is ultimate and sacred and nothing exists beyond nature. If the natural system, like the monarchial God, is all-determining and omnipotent, can genuine human freedom be defended? For Burhoe, sin is the incomplete adaptation of culture to genes, and salvation from sin is the achievement of a better adaptation. But might ethical decisions sometimes require us to resist genetic predispositions rather than adapting to them? Burhoe's view of the impersonal system of nature seems to conflict with the Christian understanding of a loving, personal God who transcends nature.[18]

Willem Drees, like Burhoe, combines a philosophy of naturalism with an appreciation of *the functions of religion* in human life, contending that "the natural world is the whole of reality that we know and interact with."[19] He acknowledges that this is a metaphysical position and not a scientific conclusion, but he says that it is a minimalist metaphysics that stays close to science. He holds that religions were adaptive in evolutionary history because they encouraged loyalty to groups broader than genetically related kin. He says that today, however, "an evolutionary view challenges religions not only by offering an account of their origins, but also by undermining the credibility of their references to a reality that transcends the environments in which the religions arose."[20] Religious traditions are adaptations to local historical environments, and as such they provide no grounds for universal claims about reality. Drees summarizes his position:

> This is unlike realism in the sense that it does not seek to protect (by reinterpretation or otherwise) the truth-claims of religious metaphors and models of an earlier age—because the metaphors and models are not understood as truth-claims but as language which was functional since it helped the individual while living his/her life and the community by creating and maintaining a culture.[21]

However, Drees does not completely exclude all ideas of transcendence. He states that *limit-questions* about the world as a whole cannot be answered by science: Why is there something rather than nothing, and why does the world have the order it has? We can express wonder and gratitude toward the reality that gave us birth. But he says that speculation about limit-questions cannot inspire devotion or worship. Drees concludes that *particular religious traditions* are still important, though he is agnostic about their claims. "Religious naturalism can build upon the heritage of religious traditions and can be open to, but at the same time be agnostic about, the idea of a nonnatural *ground of reality*."[22]

Drees's *religious naturalism* does not conflict with science (indeed, it is closely integrated with science), but it does conflict with (or remain agnostic about) most of "the heritage of religious traditions" on which he wishes to build. David Griffin says that "Drees's religious beliefs are so minimal as to be virtually non-existent."[23] I respect Drees's detailed discussion of recent literature on science and religion, which antireligious naturalists such as Atkins, Dawkins, and Wilson summarily dismiss without careful examination. But Drees is not agnostic about naturalism, which he defends as a metaphysical position rather than simply as a philosophy that might be functional in human life. With some reluctance, then, I would have to classify him, along with Burhoe, under the heading of Conflict.

INDEPENDENCE

Some authors claim that theological and scientific accounts cannot conflict because God's primary causality is on a completely different plane from secondary causes in the sphere of nature. Others maintain that science and theology are complementary languages expressing differing but not competing perspectives on reality.

1. Primary and Secondary Causality

In recent renditions of Thomistic thought, God as *primary cause* is said to work through the *secondary causes* that science studies. The two kinds of cause are inseparable, but they are on such different levels that theological and scientific accounts of events are completely independent of each other. Neo-Thomists defend the classical view of divine omnipotence and argue that it is not inconsistent with the existence of a law-abiding world or human freedom.

The Catholic scholar Étienne Gilson invokes the model of *a worker and a tool*. In God's hands, he says, "creatures are like a tool in the hands of the workman." One can say that an ax cuts the wood or that the person using the ax cuts the wood, since each produces the whole effect. Gilson insists on the reality of secondary causes. He argues that it is misguided to say that natural causes are only the occasions on which God produces the effects. God delegates causal efficacy to the creatures; they are genuine centers of activity, interrelated and dependent on each other as well as on God. Lawfulness prevails because each being has its essence, its natural way of behaving, and so it always produces the same effect.[24] How can the same effect be attributed to both *divine* and *natural* causality? Gilson says that the whole effect is produced by both divine and natural causes, but under completely different aspects. God as a primary cause is in a different order from all instrumental secondary causes.

Does such divine control preclude *chance* and *human freedom?* Chance events and human decisions are not uniquely determined by lawful causes. Gilson argues that if God were merely to calculate the future from the present, as we would have to do, God could not know the future. Since God is eternal, however, the future is present to God as it will actually occur, a single definite outcome. Being above time and having unchanging knowledge, God does not know the future as the indeterminate product of its worldly causes, but as it is specified in the eternal divine decree. Within the world, an act is uncertain before it takes place. But for God there is no "before"; for God the act has already taken place.

Neo-orthodox writers have also used the idea of primary and secondary causes to defend *divine sovereignty over nature*. Karl Barth asserts that God "rules unconditionally and irresistibly in all occurrence." Nature is God's "servant," the "instrument of his purposes." God controls, orders, and determines, for "nothing can be done except the will of God." God foreknows and also predetermines and foreordains. "The operation of this God is as sovereign as Calvinist teaching describes it. In the strictest sense it is predestinating."[25] Barth affirms both divine sovereignty and creaturely autonomy. God controls, and all creaturely determination is "wholly and utterly at the disposal of his power." The creature "goes its own way, but in fact it always finds itself on God's way." All causality in the world is completely subordinate to God. When a human hand writes with a pen, the whole action is performed by both—not part by the hand and part by the pen; Barth declares that creaturely causes, like the pen, are real but "have the part only of submission" to the divine hand that guides them.[26]

As another example, consider the discussion of *double agency* by the Anglican theologian Austin Farrer. "God's agency must actually be such as to work omnipotently on, in and through creaturely agencies, without either forcing them or competing with them." God acts through the matrix of secondary causes and is manifest only in their overall pattern. "He does not impose an order against the grain of things, but makes them follow their own bent and work out the world by being themselves."[27] We cannot say anything about *how* God acts; there are no "causal joints" between infinite and finite action and no gaps in the scientific account. So, too, the free act of a person can at the same time be ascribed to the person and to the grace of God acting in human life.

In replying to Farrer, Thomas Tracy asks how two human agents might contribute jointly to an event. They might combine their causal powers (as in pulling a rope), or one might act on behalf of the other (as an official does for a king), or one might induce the other to act (using the power of persuasion). But *a basic action* (which is not instrumental to some other goal) can be initiated by only one agent, and Tracy holds that this would apply also to God. The active powers of human agents inescapably limit God's freedom. If some of the causal work is done by natural agents, it cannot *all* be done by God. There cannot be two sufficient causes for one event.[28] Tracy maintains that God acts with other causes to shape the possibilities and the context within which human self-determination occurs. In discussing the problem of evil, Tracy argues that some human choices are contrary to God's will and cannot be ascribed to divine determination.[29] He holds that the concepts of primary and secondary causality do not provide a coherent solution to the problem of God's action in a world of scientific law and human freedom, and I am inclined to agree with him.

2. Complementary Languages

The tradition of linguistic analysis in Anglo-American philosophy holds that *diverse types of language* serve radically differing functions. Writings on the philosophy of action contend that the explanation of actions by *intentions* is very different from the explanation of effects by *causes*. An action of a human agent is a succession of activities ordered toward an end. Its unity consists in an intention to realize a goal. An action differs from a bodily movement. A given bodily movement (for example, moving my arm outward in a particular way) may represent a variety of actions (such as mailing a letter, sowing seeds, or waving to someone). Conversely, a given action may be carried out through a variety of sequences of bodily movements. An action cannot be specified, then, by any set of bodily movements, but only by its purpose or intent.[30]

Analysis in terms of *intentions* does not preclude analysis in terms of *scientific laws*. The physiologist does not need to refer to my purposes in explaining my arm movement. In addition, intentions are never directly observable. Calling a movement an action involves an interpretation of the actor's intention and often requires observation over a considerable temporal span; an action may, of course, be misinterpreted and wrongly identified. The agents of actions are embodied subjects acting through their bodies. An agent is a living body in action, not an invisible mind interacting with a visible body. Yet the agent transcends any single action and is never fully expressed in any series of actions.

Similarly, we can say that cosmic history is an action of *God as agent*. Reference to divine intentions does not exclude a scientific account of causal sequences. John Compton writes:

> We can distinguish the causal development of events from the meaning of these events viewed as God's action. Scientific analysis of physical nature and of human history has no more need of God as an explanatory factor than the physiologist needs my conscious intent to explain my bodily movements. Nor does God need to find a "gap" in nature in order to act, any more than you or I need a similar interstice in our body chemistry. Each story has a complete cast of characters, without the need for interaction with the other story, but quite compatible with it. What happens is that the

evolution of things is seen or read, in religious life—as my arm's movement is read in individual life—as part of an action, as an expression of divine purpose, in addition to its being viewed as a naturalistic process.[31]

The *intentions of an agent* are never directly observable and may be difficult to guess from events in a limited span of time. In the case of God's intentions, a religious tradition provides a vision of a wider context within which the pattern is interpreted. There is indeed a strong biblical precedent for talking about God in terms of purposes in history. Today the linguistic approach would encourage us to treat the language of divine action as an alternative to scientific language, not as a competitor with it. The cosmic drama can be interpreted as an expression of the divine purpose. God is understood to act in and through the structure and movement of nature and history.

The psychologist Fraser Watts refers to science and theology as *complementary perspectives*, as mentioned in the previous chapter. He emphasizes that they represent very different and distinct types of discourse, but they are not totally unrelated because they refer to a common world (according to the critical realism that he defends). Watts first suggests that the languages of mind and brain are different forms of discourse, though they can be correlated with each other. Going to sleep, for example, can be described by changes in the rhythms of electrical brain waves or by first-person accounts of increasing fragmentation and disorientation in time and space. Similarly, science and theology are complementary:

We can talk about what is going on in the world in purely natural terms, making no reference to God, just as we can talk about the brain in purely material terms making no reference to mind. Alternatively, we can also talk about what is going on in the world in terms of the activity of God or the movement of the Spirit. I would argue that these two perspectives on the world, the natural on the one hand and the theological or spiritual on the other, are complementary perspectives on the same reality. They describe inherently related things from different points of view, in the same way as our mind- and brain-stories about going to sleep describe the single process of going to sleep from different points of view.[32]

However, Watts rejects the total compartmentalization of theology and science that is found in some versions of the Independence thesis, including the distinction between primary and secondary causes. His position allows for significant interdisciplinary interaction despite his insistence on the contrasting perspectives and types of discourse adopted in the two fields. He holds that science places constraints on religious beliefs and may be incompatible with certain theological claims, which would allow at least some role for dialogue.

DIALOGUE

A number of theologians and scientists have used concepts from recent science to portray God's relation to the world. God can be viewed as the designer of a self-organizing process or as a communicator of information. Other writers suggest that the idea of divine self-limitation is consistent with scientific evidence and with some strands of the Christian tradition. Proponents of these models of God all reject forms of divine intervention that violate the laws of nature. God is not invoked in these models to fill particular gaps in the scientific account; on the contrary, God's role is seen as different from that of natural causes. In each case, a current scientific theory provides a conceptual parallel for a theological understanding of God's relation to nature.

1. God as Designer of a Self-Organizing Process

Until the nineteenth century, the intricate organization and effective functioning of living creatures were taken as evidence of an intelligent designer. After Darwin, the argument was reformulated: God did not create things in their present forms but designed an evolutionary process through which all living forms came into being. I indicated in Chapter 4 that the self-organization of molecules leading to life suggests considerable built-in design in biochemical affinities, molecular structures, and potential for complexity and hierarchical order. The world of molecules evidently has an inherent tendency to move toward emergent complexity, life, and consciousness.

If design is understood as a detailed preexisting plan in the mind of God, *chance* is the antithesis of *design*. But if design is identified with the general direction of growth toward complexity, life, and consciousness, then both law and chance can be part of the design. Disorder is sometimes a condition for the emergence of new forms of order, as in thermodynamic systems far from equilibrium or in the mutations of evolutionary history. We can no longer accept the clockmaker God who designed every detail of a determinate mechanism. But one option today is a revised deism in which God designed the world as *a many-leveled creative process of law and chance*. The physicist Paul Davies is an exponent of this position.[33]

According to this school of thought, a patient God endowed matter with diverse potentialities and let the world create itself. God respects the integrity of the world and allows it to be itself, without interfering with it, just as God respects human freedom and allows us to be ourselves. Human freedom requires that the world has some openness, which takes the form of chance at lower levels and choice at the human level. But responsible choice also requires enough lawfulness that we have some idea of the probable consequences of our decisions.

An attractive feature of this option is that it provides at least partial answers to the problems of *suffering* and *death*, which were such a challenge to the classical argument from design. Competition and death are intrinsic to an evolutionary process. Pain is an inescapable concomitant of greater sensitivity and awareness, and it provides a valuable warning of external dangers. My main objection to a reformulated deism is that we are left with a distant and inactive God, a far cry from the active God of the Bible who continues to be intimately involved with the world and human life.

Niels Gregersen moves beyond deism in suggesting that God *continually creates* through self-organizing systems. He describes the work of Prigogine and Kauffman on the emergence of new forms of order and complexity in nonlinear systems. He distinguishes between *structuring causes*, which limit the possibilities in a given context, and *triggering causes*, which determine particular events in that context. He proposes that God acts not by unilaterally determining particular events but as a structuring cause influencing the range of possibilities within which the creatures act:

In my view, talking about God as structuring or reconfiguring the possibility space of self-productive systems, each of its own kind, is only an option for those systems that actually display a change of the overall probability pattern. This is, however, the case in evolutionary systems. . . . From a theological perspective, such an *auto*nomous process is at the same time a *theo*nomous process if God is the stimulating power of inspiration who elicits the most fruitful possibility spaces in which the creatures try out their pathways, and who also restricts other possible possibility spaces.[34]

Gregersen suggests that such a view is consistent with the divine commands in Genesis: "Let the waters bring forth swarms of living creatures. . . . Let the earth bring forth living creatures according to their kinds. . . . Be fruitful and multiply." The authors of Genesis did not have any idea of evolutionary history, of course, but they did assign to the creatures an active role in the creative process. Gregersen notes that a God who reconfigures possibilities is not the omnipotent sovereign of the classical tradition. But he finds biblical support for the idea of divine self-limitation, which we will examine below. The Spirit represents God working within the creation, participating in and responding to the joys and suffering of the world. Such a view differs from both the monarchial God of tradition and the inactive God of deism.

2. God as Communicator of Information

In Chapter 4 we noted that information is an ordered pattern that is one among many possible sequences or states of a system (of DNA bases, alphabetical letters, auditory sounds, binary digits, or any other combinable elements). Information is communicated when another system (a living cell, reader, listener, or computer, for example) responds selectively—that is, when information is coded, transmitted, and decoded in a context of interpretation. In radio transmissions, computer networks, and biological systems, the communication of information between two points requires a physical input and an expenditure of energy. But if God is omnipresent (including presence everywhere at the microlevel), no energy is required for the communication of information. Moreover, the realization of particular outcomes among the alternative potentialities already present in the quantum world conveys information without any physical input or expenditure of energy.

John Polkinghorne proposes that God's action is *"an input of pure information."* In chaos theory an infinitesimally small energy input produces a very large change in the system. Polkinghorne suggests that in imagining God's action we might extrapolate chaos theory to the limiting case of zero energy. (This differs from quantum theory, in which there actually is *zero* energy difference between alternative potentialities, so no extrapolation is needed.) Polkinghorne holds that God's action is a nonenergetic input of information that expresses holistic patterns. God's selection among the envelope of possibilities present in chaotic processes could bring about novel structures and types of order exemplifying systemic higher-level organizing principles.[35]

As noted earlier, the biblical idea of divine Word, or *logos*, resembles the concept of information. In Greek thought, the *logos* was a universal rational principle, but biblical usage also expressed the Hebrew understanding of Word as creative power. The Word in both creation and redemption can indeed be thought of as the communication of information from God to the world. As in the case of genetic information and human language, the meaning of the message must be discerned within a wider context of interpretation. God's Word to human beings preserves their freedom because it evokes but does not compel their response.[36] I suggest, however, that the divine *logos* is not simply the communication of an impersonal message, since it is inseparable from an ongoing personal relationship. The *logos* is not a structure of abstract ideas like Plato's eternal forms, or like a computer program that exists independent of its embodiment in a particular medium or hardware system. If we believe that one of God's purposes was to create loving and responsible persons, not simply intelligent information processors, we have to draw our analogies for God's communication primarily from human life, rather than from genetic codes or computer programs.

3. God's Self-Limitation

Let us examine some critiques of the classical model of divine omnipotence, first by theologians and then by scientists and philosophers. W. H. Vanstone says that authentic love is always accompanied by *vulnerability*. In human life, inauthentic love seeks control, as when a possessive parent holds onto a child. Authentic love is precarious and brings the risk of rejection. It requires involvement rather than detachment, and this also makes a person vulnerable. The biblical God is affected by the creation, delighting in its beauty but grieved by its tragic aspects. Vanstone holds that there is no predetermined plan or assured program. Rather, there is "a vision which is discovered in its own realization." Vanstone says that evil is inescapable in the long process of creation. God must wait for the responses of nature and humanity. Nature is not just the stage for the human drama; it is the result of a labor of love and as such is worthy of our celebration and care. Vanstone extends the ancient theme of *kenosis* (divine self-limitation). In the Incarnation God set aside omnipotence, "taking the form of a servant. . . . [H]e humbled himself and became obedient unto death, even death on a cross" (Phil. 2:7–8). Vanstone holds that the life and death of Jesus reveal a God of love who participates in the world's suffering.[37]

Keith Ward rejects divine omnipotence and self-sufficiency and ascribes *reciprocity* and *temporality* to God. Creativity is inherently temporal, responsive, and contingent. God's power and knowledge are limited by the creatures' power and freedom. According to Ward this is a voluntary self-limitation because God could at any time destroy or modify the world. Chance, law, and plurality in the world produce the possibility of conflict and evil. Sentience makes pain and suffering as well as pleasure and joy possible. God chooses good and accepts evil as its concomitant. Ward says that God is neither omnipotent nor helpless but guides an evolutionary process that includes law, chance, and the emergence of novelty. God's nature and purposes are eternal and unchanging, but God's knowledge and creativity are changing.[38]

In *The Creative Suffering of God*, Paul Fiddes is sympathetic to process thought, and he draws extensively from it, though in the end he departs from it. He gives detailed critiques of ideas of God's immutability, self-sufficiency, and timelessness, and he accepts the process position concerning God's relatedness and temporality. God is with us in our suffering but is not overwhelmed or defeated by it. But Fiddes does not agree with process thought that God's involvement with the world is necessary or that God needs the world in order to be fully actualized. He maintains that God has freely chosen and accepted self-limitation for the sake of human freedom. Fiddes says that relatedness, fellowship, and community are already present within the life of the trinitarian God and do not require a world to be actualized.

Fiddes asks how *God's suffering affects us.* In Christ's death we experience judgment, but we also experience an acceptance that enables us to accept the truth about ourselves. Such costly forgiveness can have a transforming effect. But Fiddes holds that this can be better expressed through trinitarian ideas: "Process thought, then, points in a valuable way to the powerful effect which an exchange of feelings between us and a suffering God can have upon us, but I believe this insight can be carried through better with the more thoroughgoing personal analogy for God which is offered in Trinitarianism."[39]

Nancey Murphy, a philosophical theologian, and George Ellis, a theoretical physicist, have coauthored a book defending *a theology of kenosis* that accepts both divine and human self-limitation. They maintain that self-sacrificial love in the life and death of Christ is a revelation of the nature of God and also provides an ethical norm of nonviolence for the Christian today. "The proper response to a kenotic God is a kenotic relation to God and to all God's creation."[40] God seeks our free response as moral agents and does not coerce our obedience. Similarly, God respects the integrity of the created order and does not intervene coercively. Murphy and Ellis note that the evolutionary process has been slow and painful:

> The process, too, must reflect noncoercive, persuasive, painstaking love, all the way from the beginning to the end, from the least of God's creatures to the most splendid. Just as sin is a necessary byproduct of the creation of free and intelligent human beings, suffering and disorder are necessary byproducts of a noncoercive creative process that aims at the development of free and intelligent beings.[41]

Murphy and Ellis suggest that the presence of *evil* and *suffering*, which are so problematic for the classical view of divine omnipotence, is more understandable if God's power is self-limited. Pain, waste, and death are inescapable features of the evolutionary process through which free moral agents have come into being. Compared to the monarchial model, the kenotic model seems to accord better both with the biblical message of the cross and with the scientific account of evolutionary history.

INTEGRATION

There is no clear line between Dialogue and Integration, but the authors discussed below seek a more systematic synthesis of science and theology, and they are willing to accept more far-reaching reformulations of classical theological doctrines to achieve it. They propose models of God as determiner of indeterminacies and as top-down cause. We will look finally at the claim that process philosophy offers a coherent metaphysics for the integration of scientific and religious thought.

1. God as Determiner of Indeterminacies

I argued in Chapter 3 that uncertainties in the predictions made by quantum theory reflect indeterminacy in nature itself, rather than the incompleteness of current theory. In that interpretation, *a range of possibilities* is present in the world. Quantum events have necessary but not sufficient physical causes. If they are not completely determined by the relationships described by the laws of physics, their final determination might be made directly by God. What appears to be chance—which atheists take as an argument against theism—may be the very point at which God acts.

These authors who use quantum theory for theological purposes are not presenting a new version of *natural theology* in which evidence from science is used as an argument in support of theism. They are proposing ways in which a God who is accepted on other grounds (such as religious experience in a historical interpretive community) might be reconceived as acting in nature. I have called such an approach *a theology of nature* rather than a natural theology.

Those who favor such an approach believe that divine sovereignty is maintained because God *providentially controls* the events that appear to us as chance. No energy input is needed, since the alternative potentialities in a quantum state have identical energy. God does not have to intervene as a physical force pushing electrons around, but instead could actualize one of the many potentialities already present—determining, for example, the instant at which a particular radioactive atom decays.[42]

Under some conditions the effects of very small differences at the microlevel are greatly amplified in *large-scale phenomena*. In nonlinear thermodynamics and chaos theory, an infinitesimal initial change can produce dramatic changes in the larger system. Similar trigger effects can occur in evolutionary mutations and in genetic and neural systems today. Scientific research finds only law and chance, but perhaps in God's knowledge all events are foreseen and predetermined through a combination of law and particular divine action. Since God's action would be scientifically undetectable, it could be neither proved nor refuted by science. This would exclude any proof of God's action of the kind sought in natural theology, but it would not exclude the possibility of God's action affirmed on other grounds in a wider theology of nature.

If we assume that *God controls all indeterminacies*, we can preserve the traditional idea of predestination. This would be theological determinism rather than physical determinism, but in both cases nothing happens by chance. However, the problems of suffering and human freedom would remain acute. Nancey Murphy has proposed that God determines all quantum indeterminacies but arranges that lawlike regularities usually result—in order to make stable structures and scientific investigation possible and to ensure that human actions have dependable consequences so that moral choices are possible. Orderly relationships do not constrain God, since they are included in God's purposes. Murphy holds that in human life God acts both at the quantum level and at higher levels of mental activity but does it in such a way that human freedom is not violated.[43]

An alternative would be to say that while most quantum events occur by chance, God influences *certain quantum events* without violating the statistical laws of quantum physics. I discussed Robert Russell's version of this proposal at the end of Chapter 3, and a similar view has been defended by George Ellis and Thomas Tracy.[44] Their views are not in conflict with the scientific evidence. A possible objection to this model is that it assumes *bottom-up causality* within nature once God's action has occurred and thus seems to concede the reductionist's claim that the behavior of all entities is determined by their smallest parts (or lowest levels). The action would be bottom-up even if one assumed that God's intentions were directed to the larger wholes (or higher levels) affected by these quantum events. However, most of these authors also allow for God's action at higher levels, which then results in a top-down influence on lower levels, in addition to quantum effects from the bottom up.

2. God as Top-Down Cause

In Chapter 4 we explored the thesis that in the biological world activities at higher levels exert a *top-down causal influence* on low-level processes without violating low-level laws. In Chapter 5 I defended a multilevel view of persons in which mental events are held to be inseparable from neural processes but influence them causally. The idea of levels of reality can be extended if God is viewed as acting from an even higher level than interactions within nature.

Arthur Peacocke holds that God exerts a *top-down causality* on the world. In his view, God's action is a boundary condition or constraint on relationships at lower levels that does not violate lower-level laws. In general, boundary conditions may be introduced not just at the spatial or temporal boundaries of a system, but also internally through any additional specification allowed by lower-level laws. In human beings, God could influence the highest evolutionary level, that of mental activity, thereby modifying the neural networks and neurons in the brain.[45] Peacocke maintains that divine action is effected in humans down the hierarchy of natural levels, concerning which we have at least some understanding of the relationships between adjacent levels. He gives a table showing a hierarchy of academic disciplines, from the physical sciences to the humanities, which study successively higher levels, with some additional disciplines addressing interlevel questions.[46] His use of top-down causality seems to me more problematic in the case of divine action on inanimate matter; we would have to assume direct influence between the highest level and the lowest level in the absence of intermediate levels—for which there is no analogy within the natural order.

Peacocke suggests that the purposes of God are communicated through *the pattern of events* in the world. We can look on evolutionary history as the action of an agent who expresses intentions but does not follow an exact predetermined plan. Moreover, he says, God influences our memories, images, and concepts, just as our thoughts influence the activity of neurons. Peacocke maintains that Christ was a powerfully God-informed person who was a uniquely effective vehicle for God's self-expression, so that in Christ God's purposes are more clearly revealed than in nature or elsewhere in history.[47]

Philip Clayton defends top-down causality in the effects of mental events on neuronal patterns, noting that ideas can cause changes in the brain and in our behavior. Yet he fully acknowledges the dependence of mental events on physical events as understood by neuroscience. He gives a careful analysis of the emergence of higher-level phenomena from physical constituents; he calls his view *emergent monism*. He uses the mind/body relation as an analogy for God's relation to the world, which he calls *panentheism*. Whereas pantheism identifies God with the world, and theism emphasizes God's transcendence and separation from the world, panentheism holds that the world is in God, but God is more than the world. God includes the world but is not exhausted by it. This view represents a balance between immanence and transcendence. "According to the panentheism I have defended, God can act on any part of the world in a way similar to our action on our bodies. At the same time, God also transcends the world and will exist long after the physical universe has ceased to be."[48] Panentheism uses spatial metaphors to distinguish God from the world, but Clayton insists that it is not the spatial distinctions that are significant, but the contrasts between necessary versus contingent existence and perfect versus imperfect being.

Closely related to panentheism is the analogy of *the world as God's body*. Grace Jantzen starts by defending a holistic understanding of the human person as a psychosomatic unity, citing support from the Bible and recent psychology and philosophy. She rejects the classical mind/body dualism, with its devaluation of matter and the body. Jantzen thinks that the classical view of God as disembodied spirit is a product of the Christian Platonism that contrasted eternal forms with a lower realm of temporal matter and concluded that God is immutable and therefore immaterial. But a few church fathers, such as Tertullian, accepted the Stoic assertion that God is embodied, though they rejected the determinism and pantheism of Stoicism.

Jantzen acknowledges that there are significant differences between God and human persons but suggests that these can be better expressed in terms of *God's perfect embodiment* than in terms of disembodiment. We have direct awareness of our thoughts, feelings, and many events in our bodies, but much is going on in our bodies of which we are not aware (for example, the processes in our internal organs). God, by contrast, has direct and immediate knowledge of all events in the cosmos. God as omnipresent perceives from every point of view, not from a limited viewpoint as we do. With such directness, God needs no analogue of a nervous system. Again, we can directly and intentionally affect a limited range of actions of our bodies; but much that goes on, such as the beating of our hearts, is unintentional. God, however, is the universal agent for whom all events are basic actions, even if some events are more significant and revelatory than others.[49]

Though God is free of many of the limitations that the human body imposes, the presence of any body does impose limitations. But Jantzen maintains that in the case of God these are voluntary *self-limitations*. God is always embodied but has a choice about the details of embodiment, which we do not have. A universe has always existed, but its present form is a voluntary self-expression. God could eradicate the present universe and actualize something different; God could exist without this world, but not without any world. God has voluntarily given the creatures considerable independence and autonomy. Here Jantzen resembles the proponents of God's self-limitation discussed earlier, though she departs from them when she says that God and the world are "one reality." But she maintains that God transcends the world, just as we can say that a person transcends physical processes if we reject a mechanistic reductionism. She also suggests that the idea of the world as God's body would lead us to respect nature and would encourage ecological responsibility. Like Peacocke and Clayton, she draws from both the Christian tradition and current science (especially biology and psychology) in reformulating the classical view of God's relation to the world.

3. Process Theology

The process view of all entities as moments of at least rudimentary experience was presented in Chapters 4 and 5. Human experience as known from within is the starting point of this analysis; a similar interiority in much simpler forms is postulated in unified entities at lower levels (though not in inanimate objects such as stones or aggregates such as plants, which lack such unity). A unified entity at any level contributes something of its own in the way it appropriates its past, relates itself to various possibilities, and produces a novel synthesis that is not strictly deducible from the antecedents. Process thought allows for several kinds of causality, none of which is coercive or totally determining. Every new occurrence can be looked on as a present response (self-cause) to past events (efficient cause) in terms of potentialities grasped (final cause).

Whitehead ascribes the ordering of potentialities to God. God as *the primordial ground of order* structures the potential forms of relationship before they are actualized. In this function God seems to be an abstract and impersonal principle. But God is also *the ground of novelty*, presenting new possibilities among which alternatives are left open. God elicits the self-creation of individual entities, allowing for creativity as well as structure. By valuing particular potentialities to which particular creatures can respond, God influences the world without determining it. God acts by being experienced by the world, affecting the development of successive moments. But God never determines the outcome of events or violates the self-creation of any being. Every entity is the joint product of past causes, divine purposes, and the new entity's own activity.[50]

According to Whitehead, God is *influenced by events in the world*. The central categories of process philosophy (temporality, interaction, mutual relatedness) apply also to God. God is temporal in the sense that the divine experience changes in receiving from the world and in contributing to it. God's purposes and character are eternal, but God's knowledge of events changes as those events occur. God influences the creatures by being part of the data to which they respond. God is supremely sensitive to the world, supplementing its accomplishments by seeing them in relation to the infinite resources of potential forms and reflecting back to the world a specific and relevant goal.

Whitehead was primarily a philosopher, though his interests were broad. A number of theologians have used process categories in reformulating specifically Christian beliefs in the contemporary world. John Cobb and David Griffin express the dipolar character of process theism by speaking of God as *creative-responsive love*. God as *creative* is the primordial source of order and novelty, which can be identified with the biblical concept of *logos* as rational principle and divine Word. God as *responsive* is temporal and affected by the world.

The process view allows for particular divine initiatives. If God supplies distinctive possibilities to each new entity, no event is wholly an act of God, but every event is an act of God to some extent. There is thus a structural similarity between God's actions in nonhuman and human life, but there are also important differences. God's basic *modus operandi* is the same throughout, but the consequences vary widely between levels of being.[51]

In *the human sphere*, according to Cobb and Griffin, God builds on the past, including existing cultural traditions, and always depends on the free responses of individuals and communities. God loves all equally, yet that love may be revealed more decisively in one tradition or person than another. God calls all, but people respond in diverse ways. Some experiences of God's grace may be felt with exceptional power, and some individuals may have an unusual commitment to the fulfillment of God's will. Process theologians bring together God's action in nature, in religious experience, and in Christ by using a common set of concepts while recognizing the distinctive features of each. Continuing creation and redemption are presented in a single conceptual framework.

Cobb and Griffin can thus speak of *Christ as God's supreme act*. In Israel there was already a tradition of divine initiative and human response that could be carried further. Christ's message and life were rooted in this past and in God's new aims for him, and he powerfully expressed God's purposes and love. Christ can be taken as incarnation of the *logos*, the universal source of order, novelty, and creative transformation wherever they occur. In Christ we see a specific and crucial instance of a more general divine action. But Christ's free decision and faithful response were also needed to actualize God's aims for him, so the full humanity of Christ was not compromised. Christ was subject to the same conditions and limitations as other persons, but he was unique in the content of God's aims for him and in his actualization of those aims. This was not a discontinuous and coercive intrusion from outside, but the decisive instance of God's creative presence throughout the world; he is thus our clue to that wider presence. If we see Christ's life and his vision of God as revealing the nature of reality, we can be open to the power of creative transformation in our own lives.[52]

I submit that it is in the biblical idea of *the Spirit* that we find the closest parallel to the process understanding of God's presence in the world. We have seen that in the Bible the Spirit was associated with the initial creation and with the continuing creation of the creatures: "When thou sendest forth thy Spirit they are created" (Ps. 104:30). The Spirit inspired the prophets and is present in worship and prayer: "Take not thy holy Spirit from me" (Ps. 51:11). Christ received the Spirit at his baptism (Mark 1:10), and the early community received it at Pentecost (Acts 2). We can acknowledge God's particular activity in the life of Christ and in religious experience within the context of God's activity in nature. We can also believe that God as Spirit is at work in other religious traditions.

Process thought thus offers distinctive answers to each of the six problems in the classical monarchial model outlined earlier.

1. *An Evolutionary World.* Process thought is in tune with the evolutionary view of nature as a dynamic process of becoming, always changing and developing, radically temporal in character. This is an incomplete cosmos still coming into being. Evolution is a creative process whose outcome is not predictable. Reality is multileveled, with more complex levels built on simpler ones; we can understand why it had to be a very long, slow process if God's role was evocation and not control. Also fundamental to process thought is a recognition of the ecological interdependence of all entities. Here there is no dualism of soul and body and no sharp separation between the human and the nonhuman. Anthropocentrism is avoided because humanity is seen as part of the community of life, similar to other entities despite distinctive human characteristics. All creatures are intrinsically valuable because each is a center of experience, though there are enormous gradations in the complexity and intensity of experience. In addition, by balancing immanence and transcendence, process thought encourages respect for nature.

2. *Chance and Law.* Within the monarchial model, any element of genuine chance is a threat to divine control; what appears to us as chance is really determined by God. Process thought is distinctive in holding indeterminacy among its basic postulates. It affirms both order and openness in nature. Here divine purpose is understood to have unchanging goals but not a detailed eternal plan; God responds to the unpredictable.

Process thought recognizes alternative possibilities, potentialities that may or may not be realized. There are many
influences on the outcome of an event, none of them absolutely determining it.

3. *Human Freedom.* Human experience is the starting point
   from which process thought generalizes and extrapolates to
   develop a set of philosophical categories that are exemplified
   by all entities. Self-creativity occurs in the momentary present of every entity. It is not surprising, then, that process
   thought has no difficulty in defending human freedom in
   relation to both God and causes from the past. In particular,
   omnipotence and predestination are repudiated in favor of a
   God of persuasion, whose achievements in the world always
   depend on the response of other entities. Process theism
   strongly endorses our responsibility to work creatively to
   further God's purposes while recognizing human frailty and
   the constraints imposed by the biological and social structures inherited from the past. We are participants in an
   unfinished universe and in God's continuing work. God calls
   us to freedom, justice, and love. Time, history, and nature
   are to be affirmed, for it is here that God's purposes can be
   carried forward.

4. *Evil and Suffering.* Human sin can be understood as a product of human freedom and insecurity. Suffering in the human and nonhuman world is no longer a divine punishment
   for sin or an inexplicable anomaly. The capacity for pain is
   an inescapable concomitant of greater awareness and intensity of experience. Greater capacity to hurt others is a
   concomitant of the new forms of interdependence present
   at higher levels of life. In an evolutionary world, struggle,
   death, and conflicting goals are integral to the realization of
   greater value. By acknowledging the limitation of divine
   power we avoid making God responsible for particular
   forms of evil and suffering. Instead of God the judge meting
   out retributive punishment, we have God the friend, with us
   in our suffering and working with us to redeem it.

5. *"Masculine" and "Feminine" Attributes.* The classical view of
   God was heavily weighted toward what our culture thinks of
   as "masculine" virtues: power, rationality, independence, and

impassibility. By contrast, process thinkers ascribe to God what our culture takes to be "feminine" virtues: nurturance, sensitivity, interdependence, and responsiveness. These authors refer to God's tenderness, patience, and responsive love. The typical male image of control and self-sufficiency is rejected in favor of images of participation, education, and cooperation. In reacting against the monarchial model of God's power, process thinkers may sometimes seem to make God powerless, but in fact they are pointing to alternative forms of power in both God and human life. Power as control is a zero-sum game: the more one party has, the less the other can have. Power as empowerment is a positive-sum game, in which both parties can benefit. In portraying both divine and human virtues, process thought integrates masculine and feminine attributes in a new way.

6. *Interreligious Dialogue.* In contrast to the exclusivist claims of revelation in classical theism, process thought allows us to acknowledge that God's creative presence is at work at all points in nature and history. But it also allows us to speak of the particularity of divine initiatives in specific traditions and in the lives and experience of specific persons. Unlike deism, it defends the idea of God's continuing action in the world— including actions under special conditions that reveal God's purposes with exceptional depth and clarity. Such a framework would offer encouragement to the path of dialogue among world religions as an alternative to both the militancy of absolutism and the vagueness of relativism. We can accept our rootedness in a particular community and yet remain open to the experience of other communities.

In summary, I believe that Dialogue and Integration are more promising ways to bring scientific and religious insights together than either Conflict or Independence. In responding to the problems presented by the monarchial model of God, I find exciting new possibilities in the use of specific ideas in recent science to conceive of God as designer and sustainer of a self-organizing process and as communicator of information. I am sympathetic with the theme of God's self-limitation. I also admire the more systematic development of ideas of

God as determiner of quantum indeterminacies and as top-down cause. Finally, I find the concepts of process philosophy particularly helpful, but I am aware that a single coherent set of philosophical categories may not do justice to the rich diversity of human experience.

All models are limited and partial, and none gives a complete or adequate picture of reality. The world is diverse, and differing aspects of it may be better represented by one model than another. God's relation to persons differs from God's relation to impersonal objects such as stars and rocks. The pursuit of coherence must not lead us to neglect such differences. In addition, the use of diverse models can keep us from the idolatry that occurs when we take any one model of God too literally. Only in worship can we acknowledge the mystery of God and the pretensions of any system of thought claiming to have mapped out God's ways.

Notes

Introduction

1. U.S. data from Gallup Poll, Nov. 1991, see *U.S. News & World Report*, 23 Dec. 1991, p. 59; comparative data from Roper Center survey, see George Bishop, "The Religious Worldview and American Beliefs about Human Origins," *The Public Perspective* 9 (Aug./Sept. 1998): 39–44.
2. Survey of random sample from *American Men and Women of Science*, see Edward J. Larson and Larry Witham, "Scientists Are Still Keeping the Faith," *Nature* 386 (3 Apr. 1997): 435–36.
3. OCLC online library catalogue as of 15 Nov. 1998. (1990s data for 1990–1997 only.)
4. Ian G. Barbour, *Religion in an Age of Science*, Gifford Lectures, vol. 1 (San Francisco: HarperSanFrancisco, 1990), chap. 1.
5. Ian G. Barbour, *Religion and Science: Historical and Contemporary Issues* (San Francisco: HarperSanFrancisco, 1997), chap. 4.
6. John F. Haught, *Science and Religion: From Conflict to Conversation* (Mahwah, NJ: Paulist Press, 1995).
7. Ted Peters, "Theology and Natural Science," in *The Modern Theologians*, 2nd ed., ed. David Ford (Oxford: Blackwell, 1997).
8. Ian G. Barbour, *Ethics in an Age of Technology*, Gifford Lectures, vol. 2 (San Francisco: HarperSanFrancisco, 1993).
9. Willem Drees, *Religion, Science, and Naturalism* (Cambridge: Cambridge University Press, 1996), pp. 43–49.
10. J. Wentzel van Huyssteen, *Duet or Duel? Theology and Science in a Postmodern World* (Harrisburg, PA: Trinity Press International), p. 3.
11. For comments on each of these world religions, see the Index of Selected Topics in Barbour, *Religion and Science*.

Chapter One

1. Jerome J. Langford, *Galileo, Science, and the Church*, rev. ed. (Ann Arbor: University of Michigan Press, 1971); William Shea, "Galileo and the Church," in *God and Nature*, ed. David C. Lindberg and Ronald L. Numbers (Berkeley and Los Angeles: University of California Press, 1986).

2. James R. Moore, *The Post-Darwinian Controversies: A Study of the Protestant Struggle to Come to Terms with Darwin in Great Britain and America, 1870–1900* (Cambridge: Cambridge University Press, 1979).

3. Charles Darwin, letters to Asa Gray (22 May and 26 Nov. 1860), in Francis Darwin, *Life and Letters of Charles Darwin*, vol. 2 (London: John Murray, 1887), pp. 312, 378.

4. Ernst Haeckel, quoted in Loren Eiseley, *Darwin's Century* (New York: Doubleday, 1958), p. 346.

5. John W. Draper, *History of the Conflict between Science and Religion* (New York: Appleton, 1874); Andrew Dickson White, *A History of the Warfare of Science and Theology*, 2 vols. (New York: Appleton, 1896).

6. John Hedley Brooke, *Science and Religion: Some Historical Perspectives* (Cambridge: Cambridge University Press, 1991).

7. Carl Sagan, *Cosmos* (New York: Random House, 1980).

8. Carl Sagan, *Contact: A Novel* (New York: Simon & Schuster, 1985), p. 315.

9. Carl Sagan, *A Demon-Haunted World: Science as a Candle in the Dark* (New York: Ballantine, 1996).

10. Edward O. Wilson, *Sociobiology: The New Synthesis* (Cambridge, MA: Harvard University Press, 1975), p. 4.

11. Edward O. Wilson, *On Human Nature* (Cambridge, MA: Harvard University Press, 1978), chaps. 8 and 9.

12. Arthur Eddington, *The Nature of the Physical World* (Cambridge: Cambridge University Press, 1928), p. 16.

13. Henry Morris, ed., *Scientific Creationism*, 2nd ed. (El Cajun, CA: Master Books, 1985); Ronald L. Numbers, *The Creationists* (New York: Knopf, 1992).

14. George Webb, *The Evolution Controversy in America* (Lexington: University of Kentucky Press, 1994), chap. 10.

15. Philip Kitcher, *Abusing Science: The Case against Creationism* (Cambridge, MA: MIT Press, 1982).

16. Phillip Johnson, *Darwinism on Trial*, 2nd ed. (Downers Grove, IL: InterVarsity Press, 1993).

17. Michael Behe, *Darwin's Black Box* (New York: Free Press, 1998).

18. "Message of His Holiness John Paul II," in *John Paul II on Science and Religion: Reflections on the New View from Rome*, ed. Robert John Russell, William R. Stoeger, S.J., and George V. Coyne, S.J. (Vatican: Vatican Observatory, 1990), p. M13.

19. Richard H. Bube, *Putting It All Together: Seven Patterns for Relating Science and the Christian Faith* (Lanham, NY: University Press of America, 1995).

20. A good introduction is Karl Barth, *Dogmatics in Outline* (New York: Harper & Row, 1949).

21. Langdon Gilkey, *Maker of Heaven and Earth* (Garden City, NY: Doubleday, 1959) and *Creationism on Trial* (Minneapolis: Winston Press, 1985), pp. 108–16.

22. A useful summary and critique of linguistic analysis and the two-language view of science and religion is given in William H. Austin, *The Relevance of Natural Science to Theology* (London: Macmillan, 1976).

23. George Lindbeck, *The Nature of Doctrine: Religion and Theology in a Postliberal Age* (Philadelphia: Westminster Press, 1982).

24. Frederick Streng, *Understanding Religious Life*, 3rd ed. (Belmont, CA: Wadsworth, 1985).

25. Christopher Kaiser, *Creation and the History of Science* (Grand Rapids: Eerdmans, 1991).

26. I. Bernard Cohen, *Puritanism and the Rise of Modern Science: The Merton Thesis* (New Brunswick, NJ: Rutgers University Press, 1990).

27. Thomas Torrance, "God and the Contingent World," *Zygon* 14 (1979): 347. See also his *Divine and Contingent Order* (Oxford: Oxford University Press, 1981).

28. David Tracy, *Blessed Rage for Order* (New York: Seabury, 1975).

29. Ian G. Barbour, *Myths, Models, and Paradigms* (New York: Harper & Row, 1974); Sallie McFague, *Metaphorical Theology: Models of God in Religious Language* (Philadelphia: Fortress Press, 1982); Janet Soskice, *Metaphor and Religious Language* (Oxford: Clarendon Press, 1985).

30. Thomas Kuhn, *The Structure of Scientific Revolutions*, 2nd ed. (Chicago: University of Chicago Press, 1970).

31. See Ian G. Barbour, *Religion and Science: Historical and Contemporary Issues* (San Francisco: HarperSanFrancisco, 1997), chap. 5.

32. John Polkinghorne, *One World: The Interaction of Science and Theology* (Princeton, NJ: Princeton University Press, 1987), p. 64; see also his *Science and Creation* (London: SPCK, 1988).

33. Holmes Rolston, *Science and Religion: A Critical Survey* (New York: Random House, 1987).

34. Stephen Toulmin, *The Return to Cosmology* (Berkeley and Los Angeles: University of California Press, 1982), part 1.

35. Anthony J. P. Kenny, *The Five Ways of St. Thomas Aquinas' Proofs of God's Existence* (New York: Schocken Books, 1969).

36. David Hume, *Dialogues Concerning Natural Religion* [1779] (New York: Social Science Publishers, 1948).

37. William Paley, *Natural Theology* (Boston: Gould, Kendall & Lincoln, 1850).

38. Richard Swinburne, *The Existence of God* (Oxford: Clarendon Press, 1979), p. 291.

39. Stephen W. Hawking, *A Brief History of Time* (New York: Bantam Books, 1988), p. 291.

40. Freeman Dyson, *Disturbing the Universe* (New York: Harper & Row, 1979), p. 251.

41. John Barrow and Frank Tipler, *The Anthropic Cosmological Principle* (Oxford and New York: Oxford University Press, 1986).

42. John Leslie, *Universes* (London and New York: Routledge, 1989).

43. Arthur Peacocke, *Theology for a Scientific Age*, enlarged ed. (Minneapolis: Fortress Press, 1993).

44. Irene Diamond and Gloria Feman Orenstein, *Reweaving the World: The Emergence of Ecofeminism* (San Francisco: Sierra Club Books, 1990); Judith Plant, ed., *Healing the Wounds: The Promise of Ecofeminism* (Philadelphia: New Society Publishers); Carol Adams, ed., *Ecofeminism and the Sacred* (New York: Continuum, 1993).

45. Sallie McFague, *Models of God: Theology for an Ecological, Nuclear Age* (Philadelphia: Fortress Press, 1987); Rosemary Radford Ruether, *Gaia and God: An Ecofeminist Theology of Earth Healing* (San Francisco: HarperSanFrancisco, 1992).

46. For example, James A. Nash, *Loving Nature: Ecological Integrity and Christian Responsibility* (Nashville: Abingdon Press, 1991).

47. Ian G. Barbour, ed., *Earth Might Be Fair: Reflections on Ethics, Religion, and Ecology* (Englewood Cliffs, NJ: Prentice-Hall, 1972); Ian G. Barbour, *Ethics in an Age of Technology* (San Francisco: HarperSanFrancisco, 1993), chap. 3; Ian G. Barbour, "Scientific and Religious Perspectives on Sustainability," in *Christianity and Ecology: Wholeness, Respect, Justice, Sustainability*, ed. Dieter T. Hessel and Rosemary Radford Ruether (Cambridge, MA: Harvard University Center for the Study of World Religions, 1999).

48. Brian Davies, *The Thought of Thomas Aquinas* (Oxford: Oxford University Press, 1992).

49. Alfred North Whitehead, *Science and the Modern World* (New York: Macmillan, 1925); Charles Birch and John B. Cobb, Jr., *The Liberation of Life* (Cambridge: Cambridge University Press, 1981).

50. David Ray Griffin, *Religion and Scientific Naturalism* (Oxford: Oxford University Press, 1999).

51. Charles Hartshorne, *The Divine Relativity* (New Haven: Yale University Press, 1948).

52. John B. Cobb and David Ray Griffin, *Process Theology: An Introduction* (Philadelphia: Westminster Press, 1976).

Chapter Two

1. Readable and reliable introductory accounts of current theories in physical cosmology include Timothy Ferris, *The Whole Shebang: A State-of-the-Universe(s) Report* (New York: Simon & Schuster, 1997); Martin Rees, *Before the Beginning: Our Universe and Others* (Reading, MA: Addison-Wesley, 1997); George Ellis, *Before the Beginning: Cosmology Explained* (London: Boyars/Bowerdean, 1993); and John R. Gribbin, *In the Beginning: After COBE and before the Big Bang* (Boston: Little, Brown, 1993).

2. Alan Guth and Paul Steinhard, "The Inflationary Universe," *Scientific American* 250 (May 1984): 116–28; Alan Guth, *The Inflationary Universe* (Reading, MA: Addison-Wesley, 1997); Andrei Linde, "The Self-Reproducing Inflationary Universe," *Scientific American* 271 (Nov. 1994): 48–55.

3. See Steven Weinberg, *The First Three Minutes* (New York: Basic Books, 1977); James Trefil, *The Moment of Creation* (New York: Collier Books, 1983).

4. Pope Pius XII, "Modern Science and the Existence of God," *The Catholic Mind* (Mar. 1952): 182–92.

5. Robert Jastrow, *God and the Astronomers* (New York: Norton, 1978), p. 116.

6. Fred Hoyle, *Ten Faces of the Universe* (San Francisco: Freeman, 1977).

7. Selections by cosmologists and philosophers defending various interpretations of the Anthropic Principle are included in John Leslie, ed., *Physical Cosmology and Philosophy* (New York: Macmillan, 1990).

8. Peter Atkins, *Creation Revisited* (Oxford and New York: Freeman, 1992), chap. 6. See discussion of Atkins's views in Chapter 6.

9. Alan Guth, *The Inflationary Universe*.

10. Steven Weinberg, *The First Three Minutes*, p. 144.

11. Steven Weinberg, *Dreams of a Final Theory* (New York: Pantheon, 1992), p. 255.

12. Gerald L. Schroeder, *Genesis and the Big Bang: The Discovery of Harmony between Modern Science and the Bible* (New York: Bantam Books, 1990); see also his *The Science of God* (New York: Free Press, 1997), chaps. 3 and 4.

13. Hugh Ross, *Beyond the Cosmos*, 2nd ed. (Colorado Springs: Navpress, 1999); see also his *The Creator and the Cosmos* (Colorado Springs: Navpress, 1995).

14. Hugh Ross, *Beyond the Cosmos*, p. 233.

15. Langdon Gilkey, *Maker of Heaven and Earth* (Garden City, NY: Doubleday, 1959).

16. Gerard von Rad, *The Problems of the Hexateuch* (New York: McGraw-Hill, 1966), pp. 131–43; Claus Westermann, *Creation* (Philadelphia: Fortress Press, 1974); Bernhard Anderson, ed., *Creation in the Old Testament* (Philadelphia: Fortress Press, 1984).

17. See Ernan McMullin, "How Should Cosmology Relate to Theology?" in *The Sciences and Theology in the Twentieth Century*, ed. Arthur Peacocke (Notre Dame, IN: University of Notre Dame Press, 1981), pp. 19–21.

18. Edmund Jacob, *Theology of the Old Testament* (New York: Harper & Brothers, 1958), p. 139.

19. David Kelsey, "Creatio Ex Nihilo," in *Evolution and Creation*, ed. Ernan McMullin (Notre Dame, IN: University of Notre Dame Press, 1985).

20. Jon D. Levinson, *Creation and the Persistence of Evil* (San Francisco: Harper & Row, 1988).

21. Joan O'Brien and Wilfred Major, *In the Beginning: Creation Myths from Ancient Mesopotamia, Israel, and Greece* (Chico, CA: Scholars Press, 1982).

22. Frederick Streng, *Understanding Religious Life*, 3d ed. (Belmont, CA: Wadsworth, 1985); Mircea Eliade, *Myth and Reality* (New York: Harper & Row, 1963).

23. *Weekday Prayer Book* (New York: Rabbinical Assembly, 1962), p. 42.

24. See Michael Foster, "The Christian Doctrine of Creation and the Rise of Modern Science," in *Creation: The Impact of an Idea*, ed. Daniel O'Connor and Francis Oakley (New York: Scribner, 1969); Stanley L. Jaki, *The Road to Science and the Ways of God* (Chicago: University of Chicago Press, 1978).

25. Thomas F. Torrance, *Divine and Contingent Order* (Oxford: Oxford University Press, 1981).

26. Albert Einstein, *Ideas and Opinions* (London: Souvenir Press, 1973), p. 262.

27. Einstein, quoted in Robert Jastrow, *God and the Astronomers*, p. 28. See Frederick Ferré, "Einstein on Religion and Science," *American Journal of Theology and Philosophy* 1 (1980): 21–28.

28. James Trefil, *The Moment of Creation*, p. 223.

29. John Polkinghorne, *One World: The Interaction of Science and Theology* (Princeton, NJ: Princeton University Press, 1987), pp. 45, 63, 98.

30. See Robert John Russell, "Contingency in Physics and Cosmology: A Critique of the Theology of Wolfhart Pannenberg," *Zygon* 23 (1988): 23–43.

31. Stephen W. Hawking, *A Brief History of Time* (New York: Bantam Books, 1988), p. 174.

32. See C. J. Isham, "Quantum Theories of the Creation of the Universe," in *Quantum Cosmology and the Laws of Nature*, ed. Robert John Murphy, Nancey Murphy, and C. J. Isham (Rome: Vatican Observatory, and Berkeley: Center for Theology and the Natural Sciences, 1993).

33. John Gribbin and Martin Rees, *Cosmic Coincidence: Dark Matter, Mankind, and Anthropic Cosmology* (New York: Bantam Books, 1989); John Barrow and Frank Tipler, *The Anthropic Cosmological Principle* (Oxford and New York: Oxford University Press, 1986).

34. Stephen Hawking, *A Brief History of Time*, p. 121.

35. B. J. Carr and M. J. Rees, "The Anthropic Principle and the Structure of the Physical World," *Nature* 278 (1979): 605–12.

36. Stephen Hawking, quoted in John Boslough, *Stephen Hawking's Universe* (New York: Morrow, 1985), p. 121.

37. Freeman Dyson, *Disturbing the Universe* (New York: Harper & Row, 1979), p. 250.

38. Mark Worthing, *God, Creation, and Contemporary Physics* (Minneapolis: Fortress Press, 1996).

39. Ian G. Barbour, *Myths, Models, and Paradigms* (New York: Harper & Row, 1974).

40. Pierre Teilhard de Chardin, *The Phenomenon of Man* (New York: Harper & Row, 1959), pp. 226–28.

Chapter Three

1. Nontechnical accounts of quantum theory are given in Nick Herbert, *Quantum Reality: Beyond the New Physics* (New York: Doubleday, 1985), and John Polkinghorne, *The Quantum World* (London: Penguin Books, 1986). A more technical account is found in Bernard d'Espagnat, *Veiled Reality: An Analysis of Present-Day Quantum Mechanical Concepts* (New York: Addison-Wesley, 1995).

2. Albert Einstein (in a letter), quoted in M. Born, *Natural Philosophy of Cause and Chance* (Oxford: Oxford University Press, 1949). See also A. Pais, *Subtle Is the Lord* (Oxford: Oxford University Press, 1982).

3. David Bohm, *Causality and Chance in Modern Physics* (Princeton, NJ: Van Nostrand, 1957); David Bohm and B. J. Hiley, *The Undivided Universe* (London: Routledge, 1993).

4. Niels Bohr, *Atomic Theory and the Description of Nature* (Cambridge: Cambridge University Press, 1934), pp. 96–101.

5. Werner Heisenberg, *Physics and Philosophy* (New York: Harper & Row, 1958) and *Physics and Beyond* (New York: Harper & Row, 1971).

6. See Paul Davies, *God and the New Physics* (New York: Simon & Schuster, 1983), chaps. 8, 12; see also his *Other Worlds* (London: Abacus, 1982), chap. 7.

7. See Ian G. Barbour, *Religion and Science: Historical and Contemporary Issues* (San Francisco: HarperSanFrancisco, 1997), pp. 18–23.

8. Pierre Simon Laplace, *A Philosophical Essay on Probabilities*, 6th ed. (New York: Dover, 1961), p. 4.

9. Bertrand Russell, *Mysticism and Logic* (New York: Doubleday, 1957), pp. 45, 54.

10. Jacques Monod, *Chance and Necessity* (New York: Vintage Books, 1972), p. 180.

11. Jacques Monod, BBC lecture, quoted in *Beyond Chance and Necessity*, ed. John Lewis (London: Garnstone Press, 1974), p. ix.

12. See Barbour, *Religion and Science*, pp. 117–20, 168.

13. Henry Folse, *The Philosophy of Niels Bohr: The Framework of Complementarity* (New York: North Holland, 1985), p. 237.

14. Niels Bohr, *Atomic Physics and Human Knowledge* (New York: Wiley, 1958), pp. 39–41, 59–61.

15. C. A. Coulson, *Science and Christian Belief* (Chapel Hill: University of North Carolina Press, 1955), chap. 3. See also D. M. MacKay, "Complementarity in Scientific and Theological Thinking," *Zygon* 9 (1974): 225–44, and chapters by Edward McKinnon, James Loder and Jim Neidhardt, and Christopher Kaiser in *Religion and Science: History, Method, Dialogue*, ed. W. Mark Richardson and Wesley J. Wildman (New York: Routledge, 1996).

16. Ian G. Barbour, *Myths, Models, and Paradigms* (London: SCM Press, 1974), pp. 77–78.

17. James Jeans, *The Mysterious Universe* (Cambridge: Cambridge University Press, 1930), p. 186.

18. Arthur Eddington, *The Nature of the Physical World* (Cambridge: Cambridge University Press, 1928), p. 244.

19. Eugene Wigner, *Symmetries and Reflections* (Bloomington: Indiana University Press, 1967), p. 172.

20. John A. Wheeler, "Bohr, Einstein, and the Strange Lesson of the Quantum," in *Mind and Nature*, ed. Richard Elvee (San Francisco: Harper & Row, 1982); see also his "The Universe as Home for Man," *American Scientist* 62 (1974): 683–91.

21. Philip Yam, "Bringing Schrödinger's Cat to Life," *Scientific American* 276 (June 1997): 124–29; Serge Haroche, "Entanglement, Decoherence, and the Quantum/Classical Boundary," *Physics Today* 51, no. 7 (July 1998): 36–42.

22. Menas Kafatos and Robert Nadeau, *The Conscious Universe: Part and Whole in Modern Physical Theory* (New York and Berlin: Springer Verlag, 1990);

Diarmuid O'Murchin, *Quantum Theology: Spiritual Implications of the New Physics* (New York: Crossroads, 1997).

23. Mark Buchanan, "Why God Plays Dice," *New Scientist* 159 (22 Aug. 1998): 27–30. See also Tim Maudlin, *Quantum Non-Locality and Relativity: Metaphysical Implications of Modern Physics* (Oxford: Blackwell, 1994), and d'Espagnat, *Veiled Reality*.

24. John Polkinghorne, *The Quantum World* (London: Penguin Books, 1986), pp. 79, 80.

25. Jean Staune, "On the Edge of Physics," *Science and Spirit* 10, no. 1 (Apr./May 1999): 15.

26. Barbour, *Religion and Science*, pp. 175–77.

27. Gary Zukav, *The Dancing Wu Li Masters* (New York: Morrow, 1979); Ken Wilber, ed., *Mystical Writings of the World's Greatest Physicists* (Boulder: Shambala, 1984); Danah Zohar, *The Quantum Self: Human Nature and Consciousness Defined by the New Physics* (New York: Morrow, 1991).

28. Fritjof Capra, *The Tao of Physics* (New York: Bantam Books, 1997), p. 266.

29. See Barbour, *Religion and Science*, pp. 177–81.

30. David Bohm, *Wholeness and the Implicate Order* (Boston: Routledge and Kegan Paul, 1980), chap. 7; Bohm, "Religion as Wholeness and the Problem of Fragmentation," *Zygon* 20 (1985): 124–33.

31. Richard Jones, *Science and Mysticism* (Lewisburg, PA: Bucknell University Press, 1986).

32. William Pollard, *Chance and Providence* (New York: Scribner, 1958).

33. Robert John Russell, "Special Providence and Genetic Mutation: A New Defense of Theistic Evolution," in *Evolutionary and Molecular Biology: Scientific Perspectives on Divine Action*, ed. R. J. Russell, William R. Stoeger, and Francisco Ayala (Rome: Vatican Observatory, and Berkeley: Center for Theology and the Natural Sciences, 1998).

Chapter Four

1. Julian Huxley, *Evolution: The Modern Synthesis* (London: Allen & Unwin, 1942); Gaylord G. Simpson, *The Meaning of Evolution* (New Haven, CT: Yale University Press, 1949).

2. Hoimar von Ditfurth, *The Origins of Life: Evolution as Creation* (San Francisco: Harper & Row, 1982).

3. S. J. Gould and N. Eldredge, "Punctuated Equilibria," *Paleobiology* 3 (1977): 115–51; Stephen Jay Gould, "Darwinism and the Expansion of Evolutionary Theory," *Science* 216 (1982): 384.

4. G. Ledyard Stebbins and Francisco Ayala, "The Evolution of Darwinism," *Scientific American* 253 (July 1985): 72–85.

5. C. H. Waddington, *The Strategy of the Genes* (New York: Macmillan, 1975); Alister Hardy, *The Living Stream* (London: Collins, 1965), chap. 6.

6. Richard Dawkins, *The Blind Watchmaker: Why the Evidence of Evolution Reveals a Universe without Design* (New York: Norton, 1987).

7. Richard Dawkins, *River out of Eden* (New York: Basic Books, 1995), p. 133.

8. Dawkins, *The Blind Watchmaker*, pp. 13, 15.

9. Daniel Dennett, *Darwin's Dangerous Idea: Evolution and the Meanings of Life* (New York: Simon & Schuster, 1995).

10. Ibid., p. 520.

11. Ibid., p. 185.

12. *Teaching Science in a Climate of Controversy: A View from the American Scientific Affiliation* (Ipswich, MA: American Scientific Affiliation, 1986); J. P. Moreland, ed., *The Creation Hypothesis* (Downers Grove, IL: InterVarsity Press, 1994).

13. Phillip Johnson, *Darwinism on Trial*, rev. ed. (Downers Grove, IL: InterVarsity Press, 1993); *Reason in the Balance: The Case against Naturalism in Science, Law, and Education* (Downers Grove, IL: InterVarsity Press, 1995), chap. 4; *Defeating Darwinism by Opening Minds* (Downers Grove, IL: InterVarsity Press, 1997).

14. S. J. Gould, "Impeaching a Self-Appointed Judge," *Scientific American* (July 1992): 118.

15. Michael Behe, *Darwin's Black Box: The Biochemical Challenge to Evolution* (New York: Free Press, 1996), p. 39.

16. Behe, *Darwin's Black Box*, p. 231.

17. For example, Robert Dorit's review of Behe in *American Scientist* 85 (Sept./Oct. 1997): 474–75; Kenneth Miller, *Finding Darwin's God: A Scientist's Search for the Common Ground between God and Evolution* (New York: Cliff Street Books, 1999), chap. 5.

18. Edward B. Davis, "Debating Darwin: The Intelligent Design Movement," *Christian Century* 115 (15–22 July, 1998): 678–81.

19. Resolution of the Council of the National Academy of Sciences (1981), cited in *Science and Creationism: A View from the National Academy of Sciences* (Washington, DC: National Academy Press, 1984), p. 6.

20. Stephen Jay Gould, *Rocks of Ages: Science and Religion and the Fullness of Life* (New York: Ballantine, 1999), p. 6; see also his "Nonoverlapping Magesteria," *Natural History* 106 (Mar. 1997): 16–22, 60–62.

21. Stephen Jay Gould, *Rocks of Ages*, pp. 206–07.

22. Stephen Toulmin, "Metaphysical Beliefs," in *Metaphysical Beliefs*, ed. A. Macintyre (London: SCM Press, 1957).

23. William Stoeger, "Describing God's Action in the World in Light of Scientific Knowledge," in *Chaos and Complexity: Scientific Perspectives on Divine Action*, ed. Robert J. Russell, Nancey Murphy, and Arthur R. Peacocke (Rome: Vatican Observatory, and Berkeley, CA: Center for Theology and the Natural Sciences, 1995), p. 249.

24. Howard Van Till, "When Faith and Reason Cooperate," *Christian Scholar's Review* 21 (1991): 31–45; "Is Special Creationism a Heresy?" *Christian Scholar's Review* 22 (1993): 380–95.

25. Ilya Prigogine and Isabelle Stengers, *Order out of Chaos* (New York: Bantam Books, 1984).

26. Stuart Kauffman, *The Origins of Order: Self-Organization and Selection in Evolution* (New York: Oxford University Press, 1993) and *At Home in the Universe: The Search for Laws of Self-Organization and Complexity* (New York: Oxford University Press, 1995).

27. Jeremy Campbell, *Grammatical Man: Information, Entropy, Language, and Life* (London: Penguin Books, 1982).

28. Susan Oyama, *The Ontogeny of Information: Developmental Systems and Evolution* (Cambridge: Cambridge University Press, 1985).

29. Jeffrey Wicken, *Evolution, Thermodynamics, and Information* (New York: Oxford University Press, 1987), p. 177.

30. F. Ayala and T. Dobzhansky, eds., *The Problem of Reduction* (Berkeley: University of California Press, 1974); Francisco Ayala, "Reduction in Biology: A Recent Challenge," in *Evolution at the Crossroads*, ed. David Depew and Bruce Weber (Cambridge, MA: MIT Press, 1985); Ian G. Barbour, *Religion and Science: Historical and Contemporary Issues* (San Francisco: HarperSanFrancisco, 1997), pp. 230–35.

31. Niles Eldredge and Stanley Salthe, "Hierarchy and Evolution," in *Oxford Surveys of Evolutionary Biology 1984*, ed. Richard Dawkins (Oxford: Oxford University Press, 1985).

32. Michael Polanyi, "Life's Irreducible Structures," *Science* 160 (1968): 1308–12.

33. Donald Campbell, "'Downward Causation' in Hierarchically Organized Systems," in Ayala and Dobzhansky, eds., *The Problems of Reduction*.

34. Eldredge and Salthe, "Hierarchy and Evolution"; see also Stanley Salthe, *Evolving Hierarchical Systems* (New York: Columbia University Press, 1985).

35. Stephen Jay Gould, *The Panda's Thumb* (New York: Penguin Books, 1980), chap. 1.

36. Fred Hoyle and Chandra Wickramasinghe, *Evolution from Space* (London: Dent, 1981).

37. John Bowker, "Did God Create This Universe?" in *The Sciences and Theology in the Twentieth Century*, ed. Arthur Peacocke (Notre Dame, IN: University of Notre Dame Press, 1981); see also Hugh Montefiore, *The Probability of God* (London: SCM Press, 1985), chap. 10.

38. D. J. Bartholomew, *God and Chance* (London: SCM Press, 1984).

39. Arthur Peacocke, *Creation and the World of Science* (Oxford: Clarendon Press, 1979), pp. 131–38; *Theology for a Scientific Age*, rev. ed. (Minneapolis: Fortress Press, 1993), chap. 9.

40. Arthur Peacocke, *Intimations of Reality* (Notre Dame, IN: University of Notre Dame Press), p. 66.

41. For a more extended discussion of process thought, with footnoted references to other authors, see Barbour, *Religion and Science*, chap. 11.

42. John B. Cobb and David Ray Griffin, *Process Theology: An Introduction* (Philadelphia: Westminster Press, 1976).

Chapter Five

1. A good survey of the fossil record of human evolution is given in Ian Tattersall, *The Human Odyssey: Four Million Years of Human Evolution* (New York: Prentice-Hall, 1993); see also his *The Fossil Trail: How We Know What We Think We Know about Human Evolution* (New York: Oxford University Press, 1995).

2. E. S. Savage-Rumbaugh, *Kanzi: The Ape at the Brink of the Human Mind* (New York: Wiley, 1994); Terrence Deacon, *The Symbolic Species: The Co-evolution of Language and the Brain* (New York: Norton, 1997).

3. Francis Crick, *The Astonishing Hypothesis: The Scientific Search for the Soul* (New York: Scribner, 1994), p. 3.

4. Ibid., p. 252.

5. Daniel Dennett, *Consciousness Explained* (Boston: Little, Brown, 1991).

6. Daniel Dennett, *Darwin's Dangerous Idea* (New York: Simon & Schuster, 1995), pp. 81–83.

7. Dennett, *Consciousness Explained*, p. 33.

8. Edward O. Wilson, *Sociobiology: The New Synthesis* (Cambridge, MA: Harvard University Press, 1976).

9. Edward O. Wilson, *On Human Nature* (Cambridge, MA: Harvard University Press, 1978), p. 201.

10. Wilson, *On Human Nature*, p. 176.

11. Michael Ruse, *Taking Darwin Seriously* (Oxford: Blackwell, 1986), p. 253.

12. Robert Wright, "Our Cheating Hearts," *Time* 144 (15 Aug. 1994): 45–52; see also his *The Moral Animal: Evolutionary Psychology and Everyday Life* (New York: Pantheon, 1994).

13. Holmes Rolston, *Genes, Genesis, and God: Values and Their Origins in Natural and Human History* (Cambridge: Cambridge University Press, 1999).

14. J. Michael Bailey and Richard Pillard, "A Genetic Study of Male Sexual Orientation," *Archives of General Psychiatry* 48 (1991): 1089–96. See also Dean Hammer with Peter Copeland, *The Science of Desire* (New York: Simon & Schuster, 1994).

15. Swedish study cited in Lawrence Wright, "Double Mystery," *New Yorker*, 7 Aug. 1995, pp. 45–62.

16. Troy Duster, "Persistence and Continuity in Human Genetics and Social Stratification," in *Genetics: Issues of Social Justice*, ed. Ted Peters (Cleveland: Pilgrim Press, 1998).

17. Jeremy Rifkin, *Algeny* (New York: Viking, 1983); Richard Stone, "Religious Leaders Oppose Patenting Genes and Animals," *Science* 268 (26 May 1995): 1126.

18. Ted Peters, *Playing God? Genetic Determinism and Human Freedom* (New York and London: Routledge, 1997).

19. Ronald Cole-Turner, *The New Genesis: Theology and the Genetic Revolution* (Louisville: Westminster/John Knox, 1993); Ronald Cole-Turner, ed., *Human Cloning: Religious Responses* (Louisville: Westminster/John Knox, 1997).

20. H. Wheeler Robinson, *Religious Ideas of the Old Testament* (London: Gerald Duckworth, 1913), p. 83.

21. Oscar Cullmann, *Immortality of the Soul or Resurrection of the Dead?* (New York: Macmillan, 1958), p. 30.

22. Lynn de Silva, *The Problem of Self in Buddhism and Christianity* (London: Macmillan, 1979), p. 75.

23. *Interpreter's Dictionary of the Bible*, vol. 4 (Nashville: Abingdon, 1962), p. 428.

24. David Kelsey, "Human Being," in *Christian Theology*, 2nd ed., ed. Peter Hodgson and Robert King (Philadelphia: Fortress Press, 1985).

25. James Keenan, *Goodness and Rightness in St. Thomas's Summa Theologiae* (Washington, DC: Georgetown University Press, 1992).

26. John Paul II, "Message to Pontifical Academy of Sciences on Evolution," *Origins* 26 (5 Dec. 1996): 414–16.

27. Warren Penfield, *The Mystery of Mind* (Princeton, NJ: Princeton University Press, 1975).

28. Karl Popper and John Eccles, *The Self and Its Brain* (New York and Berlin: Springer International, 1977), p. 355.

29. Keith Ward, *Defending the Soul* (London: Hodder & Stoughton, 1992).

30. Malcolm Jeeves, *Mind Fields: Reflections on the Science of Mind and Brain* (Grand Rapids: Baker, 1993) and *Human Nature at the Millennium* (Grand Rapids: Baker, 1997); Donald M. MacKay, *Behind the Eye* (Oxford: Blackwell, 1991).

31. Fraser Watts, "Brain, Mind, and Soul," in *Science Meets Faith: Theology and Science in Conversation*, ed. Fraser Watts (London: SPCK, 1998); see also his "Science and Theology as Complementary Perspectives," in *Rethinking Theology and Science: Six Models for the Current Dialogue*, ed. Niels Gregersen and J. Wentzel van Huyssteen (Grand Rapids: Eerdmans, 1998).

32. See Philip Hefner, "The Evolution of the Created Co-Creator," in *Cosmos as Creation*, ed. Ted Peters (Nashville: Abingdon, 1989).

33. Marjorie Hewitt Suchocki, *The Fall to Violence: Original Sin in Relational Theology* (New York: Continuum, 1995); Robert Williams, "Sin and Evil," *Christian Theology*, 2nd ed., ed. Peter Hodgson and Robert King (Philadelphia: Fortress Press, 1985).

34. Reinhold Niebuhr, *The Nature and Destiny of Man*, vol. 1 (New York: Scribner, 1943), chaps. 7–8.

35. E. C. Blackman in *A Theological Word Book of the Bible*, ed. Alan Richardson (New York: Macmillan, 1950), p. 145.

36. Humberto Maturna and Francisco Varela, *The Tree of Knowledge: The Biological Roots of Human Understanding* (Boston: Science Library, 1987).

37. Michael Arbib, *The Metaphorical Brain 2: Neural Networks and Beyond* (New York: Wiley, 1989), chap. 2.

38. Peter Kramer, *Listening to Prozac* (New York: Viking Penguin, 1993).

39. Antonio Damasio, *Descartes' Error: Emotion, Reason, and the Human Brain* (New York: Putnam, 1994), p. 252.

40. Brevard Childs, *Biblical Theology of the Old and New Testaments* (Minneapolis: Fortress Press, 1993), chap. 7.

41. H. Richard Niebuhr, *The Responsible Self* (New York: Harper & Row, 1963).

42. Alisdair MacIntyre, *After Virtue: A Study in Moral Theory*, 2nd ed. (Notre Dame, IN: University of Notre Dame Press, 1984), chap. 15.

43. Michael Goldberg, *Theology as Narrative: A Critical Introduction* (Nashville: Abingdon, 1982).

44. Oliver Sachs, *The Man Who Mistook His Wife for a Hat* (New York: HarperCollins, 1985).

45. Leslie Brothers, *Friday's Footprint: How Society Shapes the Human Mind* (New York and London: Oxford University Press, 1997).

46. Ron Harré, ed., *The Social Construction of Emotions* (Oxford: Blackwell, 1986); K. J. Gergen and K. E. Davis, eds., *The Social Construction of the Person* (New York: Springer-Verlag, 1985).

47. Ian G. Barbour, *Myths, Models, and Paradigms* (New York: Harper & Row, 1974).

48. Allen Newell and Herbert Simon, "Computer Science as Empirical Enquiry: Symbols and Search," in *Philosophy of Artificial Intelligence*, ed. Margaret Boden (Oxford: Oxford University Press, 1990).

49. Hubert Dreyfus and Stuart Dreyfus, *What Computers Still Can't Do*, 3rd ed. (Cambridge, MA: MIT Press, 1993); see also Stan Franklin, *Artificial Minds* (Cambridge, MA: MIT Press, 1995).

50. R. A. Brooks and L. Steels, eds., *The Artificial Life Route to Artificial Intelligence: Building Embodied, Situated Agents* (Hillsdale, MI: Erlbaum, 1995); Andy Clark, *Being There: Putting Brain, Body, and World Together Again* (Cambridge, MA: MIT Press, 1997).

51. Anne Foerst, "COG: A Humanoid Robot, and the Question of *Imago Dei*," *Zygon* 33 (1998): 91–111.

52. Roger Shank, "Natural Language, Philosophy, and Artificial Intelligence," in *Philosophical Perspectives on Artificial Intelligence*, ed. M. Ringle (Brighton, UK: Harvester Press, 1979), p. 222.

53. Aaron Sloman, "Motives, Mechanisms, and Emotions," in *Philosophy of Artificial Intelligence*, ed. Margaret Boden (Oxford: Oxford University Press, 1990).

54. Rosalind Piccard, *Affective Computing* (Cambridge, MA: MIT Press, 1997).

55. John Puddefoot, *God and the Mind Machine: Computers, Artificial Intelligence, and the Human Soul* (London: SPCK, 1996).

56. Philip Hefner, "Theology's Truth and Scientific Formulation," *Zygon* 23 (1988): 270.

57. Philip Hefner, "The Evolution of the Created Co-Creator," in *Cosmos as Creation*, ed. Ted Peters (Nashville: Abingdon Press, 1989), p. 232.

58. Philip Hefner, *The Human Factor: Evolution, Culture, and Religion* (Minneapolis: Fortress Press, 1993), chap. 8. See also Lindon Eaves and Laura Gross, "Exploring the Concept of Spirit as a Model for the God-World Relationship in the Age of Genetics," *Zygon* 27 (1992): 261–85.

59. Ronald Cole-Turner, *The New Genesis*.

60. Owen Flanagan, *Consciousness Reconsidered* (Cambridge, MA: MIT Press, 1992).

61. David Chalmers, *The Conscious Mind: In Search of a Fundamental Theory* (New York and Oxford: Oxford University Press, 1996), p. 305.

62. David Chalmers, *The Conscious Mind*, p. 300.

63. David Ray Griffin, *Unsnarling the World Knot: Consciousness, Freedom, and the Mind-Body Problem* (Berkeley and Los Angeles: University of California Press, 1998).

64. Alfred North Whitehead, *Adventure of Ideas* (New York: Macmillan, 1993), p. 226.

65. Charles Hartshorne, *Reality as Social Process* (Glencoe, IL: Free Press, 1953), chap. 1; see also his *The Logic of Perfection* (LaSalle, IL: Open Court, 1962), chap. 7.

66. Margaret Hewitt Suchocki, *The End of Evil: Process Eschatology in Historical Context* (Albany: State University of New York Press, 1988), chap. 5; John Cobb and David Griffin, *Process Theology*, chap. 7.

67. John B. Cobb, "What Is the Future? A Process Perspective," in *Hope and the Future*, ed. Ewart Cousins (Philadelphia: Fortress Press, 1972).

Chapter Six

1. Étienne Gilson, *The Christian Philosophy of Thomas Aquinas* (New York: Random House, 1956).

2. Dante Alighieri, *The Paradiso*, trans. John Ciardi (New York: New American Library, 1970), canto 33.

3. Sigmund Freud, *The Future of an Illusion* (New York: Doubleday, 1957).

4. Peter Atkins, "The Limitless Power of Science," in *Nature's Imagination*, ed. John Cornwell (Oxford: Oxford University Press, 1995), p. 127.

5. Peter Atkins, *Creation Revisited* (Oxford and New York: Freeman, 1992), p. 157.

6. Atkins, "The Limitless Power of Science," pp. 129, 132.

7. Richard Dawkins, *Climbing Mount Improbable* (New York: Norton, 1996).

8. Richard Dawkins, in *Growing Up in the Universe: BBC Study Guide to the Christmas Lectures* (London: BBC Education, 1991), p. 21.

9. Richard Dawkins, "A Reply to Poole," *Science and Christian Belief* 7 (1994): 62.

10. Richard Dawkins, "Is Science a Religion?" *The Humanist* 56 (Jan./Feb. 1997): 26, 27.

11. Edward O. Wilson, *Consilience: The Unity of Knowledge* (New York: Knopf, 1998), pp. 266, 267.

12. Dale Jamieson, "Cheerleading for Science," *Issues in Science and Technology* 15, no. 1 (Fall 1998): 90.

13. Wilson, *Consilience*, p. 259.

14. Wilson, *Consilience*, p. 265.

15. Ralph Wendell Burhoe, "War, Peace, and Religion's Biocultural Evolution," *Zygon* 21 (1986): 439–72.

16. Ralph Wendell Burhoe, *Towards a Scientific Theology* (Belfast: Christian Journals, 1981), p. 21.

17. Ralph Wendell Burhoe, "The Human Prospect and 'The Lord of History,'" *Zygon* 10 (1975): 367; see also his "Natural Selection and God," *Zygon* 7 (1972): 30–63.

18. See critiques of Burhoe by Joel Haugen and Hubert Meisinger in *Zygon* 30 (1995): 553–72, 573–90, and by Karl Peters and Willem Drees in *Zygon* 33 (1998): 313–21, 489–95. See also David Breed, *Yoking Science and Religion: The Life and Thought of Ralph Wendell Burhoe* (Chicago: Zygon Books, 1992).

19. Willem Drees, *Religion, Science, and Naturalism* (Cambridge: Cambridge University Press, 1996), p. 12.

20. Drees, *Religion, Science, and Naturalism*, p. 251.

21. Willem Drees, "The Significance of Scientific Images: A Naturalistic Stance," in *Rethinking Theology and Science*, ed. Niels Gregersen and J. Wentzel van Huyssteen (Grand Rapids: Eerdmans, 1998), p. 112.

22. Willem Drees, "Should Religious Naturalists Promote a Naturalistic Religion?" *Zygon* 33 (1998): 617.

23. David Griffin, "A Richer or a Poorer Naturalism? A Critique of Willem Drees's *Religion, Science, and Naturalism*," *Zygon* 32 (1997): 595.

24. Étienne Gilson, "The Corporeal World and the Efficacy of a Second Cause," in *God's Activity in the World*, ed. Owen Thomas (Chico, CA: Scholars Press, 1983).

25. Karl Barth, *Church Dogmatics*, vol. 3, part 3 (Edinburgh: Clark, 1958), p. 148.

26. Ibid., pp. 42, 94, 106, and 133.

27. Austin Farrer, *A Science of God?* (London: Geoffrey Bles, 1966), pp. 76, 90.

28. Thomas Tracy, "Narrative Theology and the Acts of God," in *Divine Action: Studies Inspired by the Philosophical Theology of Austin Farrer*, ed. Brian Hebblethwaite and Edward Henderson (Edinburgh: Clark, 1990).

29. Thomas Tracy, "Divine Action, Created Causes, and Human Freedom," in *The God Who Acts: Philosophical and Theological Explorations*, ed. Thomas Tracy (University Park: Pennsylvania State University Press, 1994).

30. Alan White, ed., *The Philosophy of Action* (Oxford: Oxford University Press, 1968).

31. John J. Compton, "Science and God's Action in Nature," in *Earth Might Be Fair*, ed. Ian G. Barbour (Englewood Cliffs, NJ: Prentice-Hall, 1972), p. 39.

32. Fraser Watts, "Brain, Mind, and Soul," in *Science Meets Faith: Theology and Science in Conversation*, ed. Fraser Watts (London: SPCK, 1998), p. 66.

33. Paul Davies, *The Mind of God: The Scientific Basis for a Rational World* (New York: Simon & Schuster, 1992).

34. Niels Henrik Gregersen, "The Idea of Creation and the Theory of Autopoietic Processes," *Zygon* 33 (1998): 359, 360.

35. John Polkinghorne, *Reason and Reality* (Philadelphia: Trinity International Press, 1991), chap. 3; see also his *The Faith of a Physicist* (Princeton, NJ: Princeton University Press, 1994), pp. 77–78.

36. John Puddefoot, "Information Theory, Biology, and Christology," in *Religion and Science: History, Method, Dialogue*, ed. W. Mark Richardson and Wesley J. Wildman (New York: Routledge, 1996).

37. W. H. Vanstone, *Love's Endeavor, Love's Expense* (London: Dartmon, Longman, & Todd, 1977).

38. Keith Ward, *Rational Theology and the Creativity of God* (Oxford: Blackwell, 1982); see also his *God, Chance, and Necessity* (Oxford: One World Publications, 1996).

39. Paul Fiddes, *The Creative Suffering of God* (Oxford: Clarendon Press, 1988), p. 157.

40. Nancey Murphy and George Ellis, *On the Moral Nature of the Universe: Theology, Cosmology, and Ethics* (Minneapolis: Fortress Press, 1996), p. 196.

41. Murphy and Ellis, *On the Moral Nature of the Universe*, p. 247.

42. William Pollard, *Chance and Providence* (New York: Scribner, 1958).

43. Nancey Murphy, "Divine Action in the Natural Order: Buridan's Ass and Schrödinger's Cat," in *Chaos and Complexity: Scientific Perspectives on Divine Action*, ed. Robert John Russell, Nancey Murphy, and Arthur R. Peacocke

(Rome: Vatican Observatory, and Berkeley, CA: Center for Theology and the Natural Sciences, 1995).

44. See chapters by George Ellis and Thomas Tracy in *Chaos and Complexity*.

45. Arthur Peacocke, *Theology for a Scientific Age*, enlarged ed. (Minneapolis: Fortress Press, 1993), chap. 3; see also his "God's Interaction with the World," in *Chaos and Complexity*.

46. Peacocke, *Theology for a Scientific Age*, p. 217.

47. Ibid., chap. 9.

48. Philip Clayton, *God and Contemporary Science* (Grand Rapids: Eerdmans, 1997), p. 264.

49. Grace Jantzen, *God's World, God's Body* (Philadelphia: Westminster Press, 1984).

50. See Barbour, *Religion and Science: Historical and Contemporary Issues* (San Francisco: HarperSanFrancisco, 1997), chap. 11.

51. John B. Cobb and David Ray Griffin, *Process Theology: An Introduction* (Philadelphia: Westminster Press, 1976), chap. 3.

52. Cobb and Griffin, *Process Theology*, chap. 6.

Index of Names

Index of Selected Topics